To So rew

The Prelude

To So Few

The Prelude

by
Cap Parlier

SAINT GAUDENS PRESS
Wichita, Kansas & Santa Barbara, California

Saint Gaudens Press
Post Office Box 405
Solvang, CA 93464-0405

Saint Gaudens, Saint Gaudens Press
and the Winged Liberty colophon
are trademarks of Saint Gaudens Press

Copyright © 2013 Cap Parlier
This edition Copyright © 2013 Cap Parlier
All rights reserved.

Print edition ISBN: 978-0-943039-37-4

Library of Congress Catalog Number - 2014915605

Printed in the United States of America

The TO SO FEW series are works of fiction. Any reference to real people, objects, events, organizations, or locales is intended only to give the fiction a sense of reality and authenticity. Other names, characters and incidents are the products of the author's imagination and bear no relationship to past events, or persons living or deceased.

—

In accordance with the Copyright Act of 1976 [PL 94-553; 90 Stat. 2541] and the Digital Millennium Copyright Act of 1998 (DMCA) [PL 105-304; 112 Stat. 2860], the scanning, uploading, or electronic sharing of any part of this book without the permission of the publisher constitutes unlawful piracy and theft of the author's intellectual property. If you wish to use material from this book (other than for review purposes), prior written permission must be obtained by contacting the publisher at:

editorial@SaintGaudensPress.com

Thank you for your support of the author's rights.

Dedication

To all those who have gone before us, and given their last full measure of devotion to the cause of freedom and the defense of those freedoms.

See other great books available from Saint Gaudens Press
http://www.SaintGaudensPress.com

Visit Cap Parlier's Web Site at http://www.Parlier.com

Acknowledgments

To John Richard and Roger Benefiel for research assistance.

To the late Colonel Bob Clapp, USMC, for his patient tutelage on the finer points of aerial combat when Cap was a young aviator.

To my wife, spouse, partner, sponsor and cheerleader, Jeanne, who tolerated the hours, days, weeks, months and years of research, writing and discussion. She has and continues to tolerate my love of flight and the need to tell a story about the greatest event in human flight.

To my reviewers: my wife, Jeanne; John Richard; and Leta Buresh, for their patience, reflection, opinions and suggestions. I believe they made it a stronger story.

Chapter 1

> Forgetting those things which are behind, and
> reaching forth unto those things which are before,
> I press toward the mark.
> --Philippians 3:13

Sunday, 4.June.1939
Union Station
Chicago, Illinois, USA

BRIAN Arthur Drummond felt so alone in the enormous building, the largest and busiest railroad station west of the Appalachian Mountains. The ocean of people around him with unfamiliar faces and a kaleidoscope of moods, expressions and actions made the sense of aloneness even more pronounced. His decision to leave the only home he had ever known, his undoubtedly disappointed and perhaps angry parents, and newly intimate girlfriend, Rebecca 'Becky' Seward, weighed heavily on him. Brian worried most about his mentor, flight instructor, friend and *aide de l'intrigue*, Malcolm Bainbridge – the former Great War fighter ace and now aviation entrepreneur. His confusion and doubt compounded as his impressions crystallized – more people swirled around him in the magnificent place than in all of Wichita, Kansas.

The large clock with six faces hanging from the ceiling in the middle of the main room told Brian, he still had more than an hour until his train to Detroit was scheduled to leave. So far, the journey was proceeding exactly as Malcolm had helped him lay it out. The one element they had not planned was trying to sleep on his bench seat with the movement of the train, and its stops and starts that left Brian tired adding to his loneliness.

Saturday night without Rebecca. The first Saturday night in nearly five months, except for the airmeet in St. Louis, he had not seen his girlfriend. Today would be the first full day without hearing or seeing her. Brian felt this was going to be the hardest part of his effort to join the Royal Air Force – separating from his love and his family.

By now, Rebecca Seward along with his parents and grandparents must have opened their letters. Brian could only imagine how they reacted or would react, but he knew none of it would be happy as he had asked. The thought of the hurt on his mother's face, or Rebecca crying because she had been abandoned were like huge magnets pulling at him to give up his dream.

Brian had carefully written the letters. Rebecca's had taken six tries before he was remotely satisfied. His mother's letter took four attempts. After finishing his mother's letter, the others, to his father and grandparents, were easier since he had captured most of the thoughts already. The letters had been Malcolm's idea which at first Brian had thought unnecessary. Now, sitting in this distant railroad station only part way to the place and time of his destiny, Brian knew the letters were the right thing to do.

The emotions he felt and the forces driving him to England were difficult for him to describe, but he also knew he had done his best. The sense of history, the belief the impending air battle would probably be the greatest such event ever, the need to fly for a purpose, a real purpose, were all thoughts he tried to convey in his letters. Brian would not know for several months that he had accomplished his intent, nor that his dreams would not lessen the pain each member of his family and especially Rebecca felt.

". . . Detroit and points east departing on track seven, all aboard, please." The public address system boomed the words into the cavity of the building.

The word, Detroit, brought him out of his thoughts of home and family. Brian's heart raced with the possibility his daydreaming may have caused him to miss his train. He frantically looked around for the signs telling him where to go. Brian started to move one step in several directions as he wanted to make a move, to do something, he could not miss his train. Finally, he saw a man in a brown suit and fedora hat who looked as if he knew what was happening.

"Excuse me, sir."

The man looked at Brian without answering.

"Do you know where track seven is?"

Again, without answering, the man pointed to the large sign on the far wall about 50 yards away. A large number, 7, with smaller letters spelling, TRACK. Brian felt a little foolish.

"Thank you, sir," Brian said. The man nodded his head to acknowledge his words.

Brian started to move toward the number seven, then realized he forgot his suitcase and small bag. The suitcase was brand new. He bought it just a few days ago with the few dollars he had to spare after accounting for his train ticket. Retrieving the suitcase, he walked quickly to the track seven portal.

"Final call. All passengers for Kalamazoo, Detroit and points East, the Northern Star will be departing from track seven. Final call. All aboard, please."

Brian was nearly running when he passed through the portal and finally saw the train on the left side of the platform. Several people were still walking along the platform and boarding the train. There was no train on the right side, so most of the people were either going to Kalamazoo, Detroit or points

East, or saying good-bye to their friends or loved ones.

Small beads of sweat rolled down his back and chest as he sat down in an open seat. The rail car was only about one third full. Brian had chosen the seat so he did not bother any other passengers. He took a couple of deep breaths as he wiped the sweat from his face with his right hand.

As the Northern Star began to move, Brian looked out the window at the adjacent train on TRACK 6 still fascinated by the size and detail of the rail cars. He had flown several different types of airplanes, but this was his first experience with trains.

As his heart and breathing returned to normal and the light of day began to replace the darkness of the station, Brian's thoughts returned once again to the people he left behind.

―

Sunday, 4.June.1939
Wichita, Kansas, USA

THE black, '34, Ford sedan with the single chrome rimmed red light on the roof arrived at the Bainbridge house about mid-morning on this beautiful, clear Sunday. Deputy sheriff Henry Kramer stared at Malcolm as he switched off the motor, and then shook his head and looked down to retrieve something. The only other time Malcolm had seen a law enforcement official on his property was when he transported the sheriff to Lincoln to retrieve a prisoner.

Malcolm knew he was going to be the prime suspect in the disappearance of Brian Drummond and he fully expected to have to face George Drummond, Brian's father, sometime soon. Malcolm was prepared although he was not quite sure what to expect.

"Good morning, Mister Bainbridge."

"Good mornin', Henry," Malcolm responded standing on his front porch leaning against one of the supports. "What can I do for ya?"

"I think you know why I'm here."

"I s'pose."

"What do you know about the disappearance of Brian Drummond?"

"He's on 'is way ta England."

"We know that much from the letters he left. Why?"

"He wanted ta join the RAF."

"Why?"

"He's a natural pilot and the best bird hunter I've ever seen. He's good 'n he knows it. I think he felt he could contribute 'n he wanted ta be parta history."

"Whoa, Malcolm." Kramer up his hand. "What do you mean by,

contribute?"

"He's a pilot. The Brits need pilots."

"For what?"

"For the war that's comin'."

A short laugh accentuated Henry's doubt. "What war?" he asked with as much sarcasm as he could muster.

"The war Mister Churchill has been talkin' 'bout."

"That crack pot. Everyone thinks he's crazy. There's not going to be any war."

"We'll see."

"So, Churchill's notion is the history Brian wants to be a part of?"

"Yes."

"Did you help Brian?"

"How d'ya mean?"

"Did you take him somewhere?"

"No."

A clear, ornate set of chirps from mockingbird's in the trees brought a lightness to the growing tension between the men. Malcolm loved the singing of the birds in the country.

"Where is he now?"

"I dunno," Malcolm said hedging against the truth.

"Where do you think he is?"

"I can't say."

"Why not?" asked Deputy Sheriff Henry Kramer with more irritation and welling anger.

"He asked me not ta."

Deputy Kramer was nearly to his breaking point. "Well, let me tell you something, Mister Bainbridge. You may well be an accessory to kidnapping which is a capital crime or contributing to the delinquency of a minor." Kramer knew the linkage was a stretch, but he wanted some leverage to open up Malcolm Bainbridge.

"He's not a minor."

"Just barely."

"I've done nothin'."

"We've got two angry parents who don't agree with you."

Malcolm knew George and Susan Drummond would be angry with him. They lost their only son to a situation they felt had no relevance to them and exposed their child to unnecessary and undue danger. He understood their anger, but he also recognized Brian's determination and desire.

Deputy Sheriff Henry Kramer continued, "What do you have to say for yourself?"

"Nothin'."

"What do I tell the Drummond's?"

"The truth."

"And, what may that be?" asked Henry mustering up all the sarcasm he could.

"Brian has seen a higher purpose and his moment in history."

A deep, protracted laugh in conjunction with some pronounced pacing accentuated Henry's feelings. "Oh, that's ripe. Where did you get that gibberish, old man?"

Malcolm's jaw tightened with his resentment over the contempt of the younger man, who had obviously not been a veteran of the terrible carnage in France. The veteran wanted this deputy to go away. Malcolm knew he was near the edge – the edge of control – and was about to say something he knew he would regret. "Ya wouldn't understand," was his simple answer.

"Try me."

"We must all stand against the tyranny a Hitler."

"Hitler," shouted Kramer, "he's Europe's savior."

Now, Malcolm knew he did not want to talk to this man anymore. His jaw tightened further and both fists clenched behind his back like hard stones.

Kramer continued without encouragement. "Hitler has been the only man who has brought prosperity and order to the mess in Europe." The deputy waited for some response. He received none. "Can't you see it?"

Still no response from Malcolm Bainbridge. Now, the old aviator knew this man was trouble.

"Well, I don't give a damn about your politics, or Churchill's. I'm just glad he's not calling the shots in England, or there would be war in Europe."

The day had been as he had predicted, but the words with this ignorant deputy were not what he had imagined, nor what he cared for. Malcolm recognized the sentiment expressed by the younger man. His voice was probably near the majority in America as far as Malcolm could determine. Mister Malcolm Bainbridge, former pilot officer in the Royal Flying Corps' No.43 'the Fighting Cocks' Squadron, was convinced he was right. Mister Churchill was right. He was also convinced more than ever, Brian was right.

"Enough of this," Kramer said referring loosely to the political words. "Why did Brian want to join the Brits?"

". . . defend freedom against the forces a evil."

Kramer sneered. "Ha. Did you give him that crack pot idea?"

"I've had enough a yar foolishness. If ya don't have any other questions, I'll ask ya ta leave."

The sheriff ignored Malcolm. "So, he's off to England, then?"

"Yes."

"That means he's catching a ship out of New York or some port, right?"

"I don't have ta answer that."

"You'd damn well better, old man, or I will arrest you."

"Really," challenged Malcolm. "For what?"

"Accessory to kidnapping or obstruction of justice."

A broad smile illuminated Malcolm's face. "I'm afraid ya've missed there. No laws have been broken."

"Yeah, well, I'll think of something if you don't tell us where he's catching the boat."

"Ya'd better leave," Malcolm said as he turned to return to the house.

"Maybe so old man," the sheriff responded to his back, "but I'll probably be back until we find young Master Drummond." The deputy did not wait for a response as he turned to leave. Deputy Sheriff Henry Kramer opened his car door and shouted back at Malcolm, "I think I'll just notify the FBI. Maybe it isn't kidnapping, but there is a federal law about US citizens joining the armed forces of some other God damned country. You won't like how this comes out, you smart ass old man."

Malcolm chose not to respond from the sanctuary of his home. Kramer chose not to press the point further. Malcolm was glad to see the man depart, but he also suspected it was not the last he would see of Deputy Sheriff Henry Kramer.

―

Sunday, 4.June.1939
Aboard the Northern Star in southern Michigan

THE rich greens of the fields and trees along the tracks blended with spurts and fits of sleep stolen on the bench seat with his head against the window or the seat back. Brian tried to pay attention to the terrain and vegetation of this area of the country. The only appreciable difference between Illinois and Michigan was the trees. There were simply more trees, not different, just more of them. The buildings were also different. They were closer together as if the space between the buildings was less important than it was in Kansas. Of course, the tall buildings of Chicago were markedly unique. Brian had never seen anything like them.

Only three apples remained in the small bag he carried with his suitcase. Malcolm had told him the food on the trains was pretty expensive, and he had been right. There was not a great deal of money left to spare.

Although Brian had a long way to go before arriving in England, he had no way to know what to expect after he reached Windsor, Ontario, Canada. Malcolm had told him what would probably happen, but there was more than one way to travel to England. The most likely was by ship, maybe a freighter or maybe even a passenger ship, according to Malcolm. There was also the occasional flight by Pan American Airways. He was sure there were other airlines, but he did not know the names.

The plan Malcolm helped him with also told him the next stop was going to be his greatest obstacle. The problem was most likely to be the police or the border customs agents. It would be not quite two days since he left Wichita and his parents would probably have the police looking for him. It was agreed between mentor and protégé to tell no lies. Malcolm said he would do his best to distract them from his trail, but they would probably figure out the plan fairly quickly. A few telephone calls to key locations could prevent Brian from crossing the border. As a young man on his way to join the Royal Air Force, the authorities could arrest him. It was questionable whether they would return him to Wichita regardless of what he wanted to do. Malcolm had given him a few good hints to evaluate the situation with the police in Detroit and the Customs Agents at the border checkpoint. Malcolm had told him, he would probably make it across if he was careful and attentive.

As his mind floated between his planning angst and those he left behind and the smear of interlaced trees and farm fields, an odd but familiar movement caught his eye. A radial engine biplane skipped very low across the trees and field toward the train . . . nearly directly toward him. Brian rose in his seat as the plane approached. The deep, rapid thump of the engine could be heard above the monotonous clatter of the rails and base noise of train. Brian shot to an open seat on the opposite of the train to watch the swiftly retreating airplane.

"You like airplanes, do you lad?" came a deep, male voice behind him.

Brian nodded his head until the aircraft was out of sight. He turned to see the brown tunic, single silver bars on the epaulets, and the silver, winged shield that designed the man as a Army Air Corps pilot. "Yes, I do, sir."

"Great machines."

"I know."

"Oh really, how so?"

"I am a pilot also."

The lieutenant sat down next to Brian, stuck his right hand and said, "My name is Johnson, Jay Johnson."

Brian grasped his proffered hand firmly. "Brian Drummond."

"So, how many hours do you have?" he asked as if to test the veracity of his claim.

"I don't know precisely, but I would guess eight hundred."

The lieutenant shook his head. "Now, how could that be? You're just a kid."

Brian looked him in the eyes. "I have been flying since I was nine."

"Damn, boy. I have only been flying for a year and barely have a hundred hours."

Brian shrugged his shoulders.

"Where you headed?"

Brian searched his eyes, and then looked out the window at a large field dotted with Jersey cows.

"Are you running away from home?"

"No," Brian answered to the window. He remembered Malcolm's words of caution to avoid disclosure of his intentions. "I am eighteen," as if to say he could make up his own mind.

"So, you must be headed off to fly against the wishes of your family."

"I'm not looking for trouble, sir."

"And, I'm not fixin' to make any for you. I'm just looking for conversation with a fellow pilot."

'Conversation with a fellow pilot' had a certain ring to it for Brian. He wanted to talk to someone. Why not a fellow pilot? "I'm heading to Canada to join the Royal Air Force."

The surprise exploded across Johnson's face. "Damn, boy!" he said looking around for inopportune listeners. He lowered his voice, "do you know what you're doing?"

"Yes."

"You've heard of the Neutrality Act, haven't you?"

"Yes."

"Then you must really have a fire in your belly."

"Yes, sir. I do."

They talked in hush tones with occasional flares of excitement about flying, the looming clouds of war, and the prospects of fighter pilots with faster, more agile, and lethal aircraft. Brian told him of Malcolm Bainbridge and Malcolm's friend and colleague, Royal Air Force Group Captain John Spencer – staff officer in Headquarters Fighter Command and cousin to the famous Winston Churchill. Johnson was impressed and supportive.

The train began to slow. "This is my stop," Johnson said. "Yours will be the next one."

"Thank you, sir."

"Thank you, Brian. Good luck. You're going to need it where you're headed."

"Thank you, sir."

Johnson stood and gathered his bag and small briefcase. "Who knows, perhaps we will meet again."

"Yes, sir."

They said good-bye, and Brian was alone, again. The train jerked to a stop at a small station. More people got off than got on. They waited at the station for fifteen minutes before the journey continued.

According to the train schedule, the Northern Star was about two hours out of Detroit's Grand Central Station. It would be late afternoon on Sunday when he arrived. Brian thought about spending the night in Detroit for two good reasons as far as he was concerned. He needed a good night's sleep and a shower. It was also going to be his last night in the United States of America for probably a long time. The thought passed with the image of the Detroit police stopping him from crossing the border. The more time he gave the police to catch up to him, the more likely they would be able to stop him. No, rest was not the best thing to do right now. The immediate objective was to cross the bridge over the Detroit River and see the Canadian flag, and then he would find a place to rest.

Brian's thoughts drifted toward his ultimate objective. He remembered the details of Malcolm's description of the new Supermarine Spitfire fighter airplane. The idea of flying such a powerful and fast aircraft was like a strong narcotic to an addict. He needed the energy of one of the best airplanes in the world.

The signs of fatigue faded although Brian had only managed to steal occasional catnaps since Saturday morning. With his approaching arrival in Detroit, all his thoughts focused on the actions he needed to take. Malcolm felt his best chance would be to get out of the train station quickly and hire a taxi to drive him to the border crossing. It was an expense that he did not need, but Malcolm felt it would provide less exposure, and make him seem older and more self-sufficient than he really was. Movement from the train station to the far side of the Detroit River should be without hesitation or delay. Brian needed to look as if he was a man with a mission, which he was.

Sunday, 4.June.1939
Wichita, Kansas, USA

    MALCOLM Bainbridge wanted to get away from the maelstrom building around him. He knew the deputy was right. The law would be back. He also knew he would have to face George and Susan Drummond, Brian's parents. It was essentially a foregone conclusion; they were not going to be happy with him. Malcolm had confidence he would be able to deal with the adults. What he was not sure about was trying to answer Rebecca Seward's questions and dealing with her anger that would surely be directed at him.

    The attraction of the airplanes was strong. Extrication from the gathering whirlpool sucking him in would be easy. Simply jump into one of his planes and takeoff for anywhere. Even flying overhead would be better than having to face all the people who did not understand why Brian did what he did, or why he wanted to do it.

    Retreating from the confrontations would not help Brian nor his family and girlfriend. Malcolm knew he had to face them and try his best to help them understand. If he was successful, they would feel better, and they would be better able to support Brian in what was undoubtedly going to be a very difficult time. Brian needed that. It could very well be the most valuable assistance he gave to Brian Drummond.

    The late afternoon sun was warm, verging on hot, with the characteristic wind of the Great Plains summer that seemed to be coming early again this year. Malcolm's hand moved over the skin of the Sopwith F.1 Camel like a trainer might feel the muscles of a prize thoroughbred horse. The airplanes and the prospect of losing his worldly concerns in flight were not enough. He still had a bad taste in his mouth from the morning's questioning by the obnoxious deputy. Something more had happened this morning. The troubled thoughts rolling through his mind were soothed somewhat by the rustling of the leaves in the wind and the melodious birds.

    Had he done the right thing helping Brian? Had he assisted a bright, enthusiastic young man in committing figurative suicide sending him off to a certain brutal war? Was it his stories of France that gave Brian the idea of flying for England as he did? Why had Brian really wanted to go over there? Was it the flying? Was it the adventure?

    There were no answers, just more questions. Malcolm refused to consider the consequences of something unpleasant happening to Brian. His sixth sense told him everything would be OK. Right now, he had to believe.

    Malcolm's thoughts drifted off to points north. If everything was going according to plan, his young protégé should be in Detroit and crossing the border into Canada. There was some satisfaction that John Spencer would take care of Brian from that point onward. Once in Canada, the young man would not have to worry about anything except winning each confrontation in the skies over Europe.

The contemplation of what is and what might be was brought to an abrupt end as a '36 Ford flatbed truck drove up to the house. Before the vehicle stopped, George Drummond jumped out. Spotting Malcolm by the barn hangar, the smaller, younger-looking man moved quickly toward him. The clenched fists were not a good sign.

George Drummond started shouting when he was still four yards away. "What have you done to my son?"

Malcolm held up both hands as if he was surrendering to arrest for a crime. There were no words that came to his mind.

"Why have you filled his head with such foolishness?"

Again, no answer, but now George was closing to within striking distance. Malcolm made no move to defend himself sensing that any action would only antagonize George Drummond.

"Answer me, God damn it. What have you done to my son?"

"I haven't dunna thing ta ya son."

"Then, where is he?"

"He's probably in Canada by now."

"Canada," screamed George Drummond as his body twitched on the verge of striking Malcolm. "You son of a bitch."

Restraint was still the watchword as the urge shot forward to at least fend off any blows that might come. Somehow Brian's father deserves to vent his anger, Malcolm told himself. The venerable aviator and Great War ace tried desperately to remain absolutely still, allowing not even a tightening of his jaw.

"Why? Why have you filled my son's head with all this nonsense? Airplanes. War. The Royal Air Force, for God's sake. Why have you taken my son?"

It was time to answer the accusations. "Mister Drummond, I haven't taken . . ."

"You have to, you son of a bitch. You filled his head with all this crap," George shouted waving his hands toward Malcolm's airplanes. "You are an accomplice to kidnapping."

"Wait," Malcolm spoke softly holding his hands up like a traffic cop. "If ya'll permit me, I'll try ta he'p ya understand."

"I don't want to understand. I want my son back."

"Please let me explain."

"I'm all ears. I'm angry as hell, but I'm all ears."

"Wouldja care ta sit," Malcolm offered moving his open right hand toward the porch.

"No!"

"Mister Drummond, I'm truly sawry for yar perceived loss. Y'all haven't lost yar son. He's gotta dream. He's a natural pilot. He loves ta fly. He's the best damn natural, instinctive pilot I've ever known, and I've seen a good many including Eddie Rickenbacker."

"Why England?" asked George Drummond beginning to calm somewhat.

"I'm not sure I know," Malcolm said holding back his true feeling. He knew exactly why Brian wanted to go to England. It was precisely the same reason he joined the Royal Flying Corps 23 years earlier.

"Deputy Kramer tells me, you believe there is going to be another war in Europe. Is that true?"

"Yep."

"How can you say that when the British and Germans have an agreement?"

"I believe Churchill's right. There's goin' ta be a war. Hitler keeps takin' everythin' they feed 'im. Hell, that madman's taken all of Czechoslovakia, for God's sake. Chamberlain's damn 'peace in our time' Munich Accord is just a scrap a paper."

"Why did you encourage him to go?"

"Mister Drummond, ya gotta believe me. I tried ta talk 'im outta the notion."

"How can I get my son back?" asked Brian's father as reality began to sink in.

"I wouldn't suggest that."

"How can I get him back?" snapped George.

"I can try ta get an address for 'im once he arrives in England."

"When?" the worried parent asked with an even more subdued voice.

"I would guess he'll get there in a few weeks or so. It'll take a few weeks for me ta get an address."

Regaining some strength, George Drummond said, "He's our only son, our only child. We want him back safely. I still hold you responsible for giving him these crazy ideas. I want to know an address as soon as possible, you understand?"

"Sure. I understand."

"Thank you, Mister Bainbridge. We'd appreciate all the help you can give us to get our son back."

"Sure."

George Drummond nodded his head and left. Eventually, he knew he would also have to talk to Rebecca Seward as well as others. This confrontation was not over. The thought of possible recrimination if Brian was injured or killed flashed quickly through his mind.

Leaning his head through the front door, Malcolm shouted, "Gert, hon, I've gotta go fly. I needa break."

"OK, Malcolm. Please be safe."

The tensions of the day faded as rapidly as the ground fell away. Malcolm looked for a small cumulus cloud. The airplane shook and shuddered as he passed through the little cloud several times flirting with the periphery and penetrating the interior. The experience ahead of young Brian Drummond seemed to dominate his thoughts even with the distraction of flight. He gained a clear appreciation for the agony that consumed his parents when he had volunteered to go to France. It was not a good feeling.

—

Chapter 2

> It behooves every man who values liberty of conscience for himself,
> to resist invasions of it in the case of others; or
> their case may, by change of circumstances, become his own.
>
> -- Thomas Jefferson

Sunday, 4.June.1939
Grand Central Station
Detroit, Michigan, USA

THE dual track split into four, then eight. Before Brian Drummond realized the change, there were more tracks than he could count. If the conductor had not just announced their arrival at Grand Central Station, the scene outside could have easily been Chicago, again. It looked the same except for the rain and dark gray clouds.

The dim light quickly became quite dark as the train slowly advanced into the caverns of the station. People in his railcar began to rise gathering their coats, hats and bags anticipating their arrival. Several children ranging from about 3 years old to teenagers moved more quickly around the car darting between adults despite the meager protestations of their parents. They were excited the confining train ride was finally over.

The lights of the platform began to appear. Brian's heart beat slightly faster, and he could feel the low level adrenaline produced changes in his body.

"This is it," he said aloud.

"Excuse?" an elderly woman asked thinking Brian was talking to her.

"Nothing, ma'am, I was just talking to myself."

"Then you should find someone to talk to. Words spoken for no purpose are words wasted."

"Yes, ma'am," Brian responded as he began to concentrate on the people he could see on the platform.

No policeman, yet. The cautions Malcolm had given him were reviewed several times as the train crept to a stop. His principal choices were to depart with the crowd making their detection of him more difficult, if they were looking for him. Malcolm had told him the crowd choice was a risk because he was taller than average making it somewhat easier to spot his youthful face. The other choice was to go out the other side of the train across the adjacent tracks to not be a part of the arriving passengers from Chicago. If he was spotted going across the track to the other platform, he would most certainly draw attention to himself as someone who needed to hide.

Malcolm had also told him, he had a number of factors on his side.

If the police were looking for him, they would only have a verbal description. Passing a photograph by mail to Detroit would take several weeks. It was decision time.

Brian started to say aloud, then remembering the old lady's words, and said to himself, I'll take the crowd.

Out on the platform the mass of passengers moved up the platform to the station proper. There was another train on the companion track; however, it was empty awaiting its next passenger load.

Brian carefully looked up and down the platform trying as hard as he could to look natural as a curious newcomer to Detroit. His heart rate jumped noticeably when he saw the tall, burly man in the dark blue suit wearing a large silver badge over his left breast. He stood in the middle of the platform looking directly down the platform. The policeman did not move and was obviously concentrating on the activity on the platform.

Was he looking for a teenage runaway by the name of Brian Drummond? Or, was he looking for someone else? Maybe he was simply watching the passengers so everything remained orderly and peaceful. Brian stopped for a moment feeling like a caged animal desperately wanting to get out. Should he go back to plan B? Would that action draw even more attention to himself insuring detection? Malcolm had told him to make his decision and make the best of it. Anonymity was on his side.

Working his way gradually toward the edge of the platform would enable Brian to pass by the policeman with the greatest distance, only about ten feet, between them. Several people walking out also would help shield him a little.

Trying as hard as he could to look straight ahead and not looking at the policeman, Brian passed by losing the policeman from his peripheral vision. Don't look at the officer. Don't look back, just keep walking normally. Don't run, or walk too fast. Brian was burning with a tremendous urge to break into a full sprint for the doors, but he fought to contain the feeling.

The main hall was not as big as Chicago's Union Station, but it was still big. The large room was busy although it was not crowded. People moved in almost every direction. There were several exits to the gray, and still rainy exterior. Brian picked the most popular one and moved among the variety of people toward the exit.

Another police officer stood near the right side of the doorway. He had not seen the man earlier. Brian stopped to scan the other exits. There were also policemen watching those exits as well. Another obstacle to be overcome was about all Brian could think about. This obstacle was no different from the previous one.

His suit, white shirt and tie gave Brian some slight comfort. Malcolm helped him pick out the clothes, so he would appear more like a businessman and less like a boy on the run. Pretending to check his pockets for something gave Brian sometime to consider his next move. The same technique would have to work. There was a trash barrel near the left side of the exit. Brian made his way through the crowd to throw a piece of paper in the barrel, and then walked out the exit as far from the policeman as possible.

He was outside under a large overhang. It was still raining although not very heavily. It was more like a drizzle than rain.

Quickly, Brian looked to an available taxi. The little sign on the top which read, TAXI, was the best indication, but they were also painted with strips of yellow squares that looked like a band of yellow and black checkerboard around the black car. Several were filling up. Another Ford Sedan taxi drove up letting passengers out. Brian moved toward the new arrival.

A hand grabbed his shoulder. Turning quickly to see who it was, his heart skipped a few beats as he recognized the uniform of the Detroit Police Department. Brian thought about trying to break and run thinking he might be able to outrun the older man.

"Excuse me, son," he said with a heavy Irish accent. The man looked directly into Brian's eyes. "I didn't mean to startle you." Holding a small penknife in his open hand, he continued, "I believe you dropped this."

Brian looked quickly from the officer's face to the knife and back. It was his penknife. "Thank you, sir," said Brian taking the proffered knife.

"You're welcome. You need to slow down a wee bit, me boy."

"Yes, sir."

Nodding to Brian, he concluded the short conversation. "Have a good day."

"Thank you, sir."

Brian turned as slowly as he thought he could, which was actually quite fast. The taxi was still empty. Leaning into the cab, Brian asked as softly as he could so the policeman, if he was still there, could not hear, "Can you take me to Canada?"

"Sure, sonny. Hop in."

Situating his bag on the seat beside him, Brian closed the door behind him. The cab driver was a large, round man with a bright red face. His whole body looked as if it was going to explode. He was almost entirely bald or at least that was his appearance since he was wearing a strange sort of five pointed hat.

"Where to, sonny?"

Why did everyone keep referring to him as a boy? Did he really look that young? Maybe it was just a form of greeting people in Detroit used when addressing someone younger? Brian hoped he did not look too obvious. All those policemen in the train station must have been normal although he had never seen so many police officers at one time before.

"Canada, please."

"You already said that and Canada is a big country. Where to in Canada?"

"Windsor."

"OK, that's just across the river. Now, where to in Windsor?"

The questions had come a little too quickly for Brian. He hesitated struggling to remember the street names of his destination. "Maple and Lewis Streets."

"That's in the business district. There's nothing down there on a Sunday evening."

Brian was not sure if he was supposed to have a response or provide other directions. He figured he would wait for a question. He did not have to wait long.

"So, where do you want to go?"

Feeling some frustration, Brian said, "I just want to go to Maple and Lewis Streets in Windsor, Canada."

"OK, but it's going to be pretty quiet."

They started to move. Brian took the opportunity to look back at the station to see if the policeman was watching him. There were no dark blue, police uniforms in sight. Brian sank back into the seat and exhaled a deep breath.

After several blocks passed as a blur to Brian, the taxi driver spoke. "Who you running from young man?"

"Why do you ask that?"

"I can tell. I've seen the look too many times."

"I'm not running from anyone."

"Sure. So, then, why are you going to Canada?"

Malcolm had told him to expect the question at the border. "I'm going to visit some friends of my family."

"Sure you are, son. That's why you asked to go the business district where no people live and on a rainy Sunday evening, ta boot."

Brian recognized the disconnect immediately. His brain slammed into overdrive trying to think of a story that would tie all the obvious facts together. The sense of desperation returned as the fabrication of a reasonable story eluded young Brian Drummond.

"Look boy, I'm not the cops. I'm not going to turn you in. I just want to know what I'm getting into before I get into it."

There was genuineness to the man's words. Brian decided for some reason to trust him. "I'm going to Canada to join the Royal Air Force."

"I'll be damned. I've heard about this, but never met someone doing it. Just a couple of weeks ago, the FBI picked up a guy trying to do the same thing. I'll be damned."

Brian was not quite sure what the man was referring to, but the words, 'FBI' and 'trying to do the same thing,' sent a shiver through his body. It was exactly what Malcolm described about the federal law that presented his most serious obstacle. He was also not particularly eager to divulge too much even if he did trust the man.

"Are you a pilot?"

"Yes."

"You're too young to be a pilot."

"That maybe, but I'm a pilot nonetheless."

They continued to move through the nearly deserted downtown streets. "Why are you joining the Royal Air Force?"

"Because there's a war coming, and they need my skills."

"I'll be damned," he said pausing to consider the words. "Why do you think there's going to be another war?"

"Because a very good friend has a friend who's uncle is Winston Churchill."

"So."

"So, Winston Churchill believes there is going to be a war."

"Is that so?"

"Yes."

Another block of buildings passed before the next question came. "How old are you, son?"

"I don't think that is any of your business," Brian said his rehearsed response with a little resentment to his voice.

The taxi driver pulled over to the curb and stopped the car. Turning in his seat to place his huge right arm on the seat back, the man looked directly into Brian's eyes with his bulging, large brown eyes. "You listen here. It is against the law for me to transport an underage person across the border. So, I've got to know how old you are."

"How old do you have to be?"

"Twenty one."

"Then, I'm twenty one."

"My ass, you are," the man said with anger coming to his voice. "You either tell me how old you are, or you get out of this cab. You got it?"

"Yes."

"So, let's try this again. How old are you?"

Brian considered the alternatives. Malcolm had told him the best chance he had to get across the border was in a taxi. "I'm eighteen."

"Eighteen. Jesus H. Christ. What the hell are you doing, boy? Are you crazy? The Brits aren't going to take a boy into the RAF."

"They said if I could get to Canada, they would take me."

"Oh right. And, who is they."

"Group Captain John Spencer of the Royal Air Force."

"Where you from, boy?"

"Wichita, Kansas."

The answer brought a heavy almost choking laugh to the rotund taxi driver. "A farm boy in the RAF, now I've heard everything. So, you want me to take you across the border to join the RAF."

"Yes."

The man turned back around and stared down the street for what seemed like a eternity to Brian. He started to drive away, but stopped abruptly. Turning back to Brian, he asked, "Do you have your passport?"

"Yes."

"Let me see it, please."

Brian dug his newly acquired passport out of his small bag and handed it to the taxi driver.

"Oh shit," the man said with obvious disgust. "You put your real age on here."

Remembering the sequence as outlined by Malcolm helped Brian. "I have a letter giving me permission," Brian said retrieving the paper Malcolm had written and handing it to the driver.

The man looked directly at Brian after reading the letter. "This is not from your father, is it?"

The thought of divulging the source of his permission made Brian feel extremely vulnerable. But, he had trusted this taxi driver until now. "No. He's my friend."

"And, whom, may I ask, is your friend."

"Malcolm Bainbridge. He was a Pilot Officer in the Royal Flying Corps during the Great War."

"Ah, ha. Now, I'm beginning to see the picture. So, you're bound and determined to do this thing even at your young age because your hero did it?"

"No . . . but yes . . . I guess."

"Well, then, let's get on with it and see if we can get you across the border to meet your dream."

The remaining drive through the city did not take very long. The large body of water directly ahead of them gave Brian a strange mixture of emotions. He felt excitement and elation that he could practically see the most important waypoint, and probably greatest obstacle, of his journey. And yet, he also felt considerable apprehension over the challenge before him.

Brian and Malcolm had agreed to this part of the plan with all its risks. His best chance to get across the border was to be forthright about his age, since he looked young, and to camouflage his motive for crossing the border. At least, the taxi driver thought he could make it, or they would not still be heading toward the river. The acceptance by the taxi driver gave Brian a little shot of confidence that he truly needed.

As they approached the river, the large arc of the Ambassador Bridge connecting Detroit with Windsor became the dominant feature. Brian could see the customs check point on the U.S. side. He could not see the Canadian check point on the far side because of the bow in the bridge roadway.

"Now," said the driver, "one piece of advice before we hit the customs gate. Don't volunteer any information. Answer their questions with as short an answer as possible."

"Yes, sir."

"For God's sake, don't tell them you're joining the RAF?"

"Yes, sir."

"Do you understand?"

"Yes, thanks."

Several other automobiles and small trucks were waiting at the gate in each direction. The taxi pulled up behind the automobile in the right lane. Brian watched the customs agents intently trying to determine if they were looking for him or anyone else. The agent in front was talking to the driver of the car and looking into the interior as if he was trying to identify the other people. In the left lane, the contents of the truck were being inspected. Brian tried to imagine what the agent was inspecting for, but it was not obvious. The car in front of them pulled away across the bridge. They had been cleared.

"Be calm, kid," the taxi driver cautioned.

Brian nodded his head even though the driver was not looking at him. The agent motioned the taxi forward. Brian concentrated on the image of himself flying a Spitfire at high speed through a cloud filled sky. The same concentration and focus he used when he flew would serve him well now. The image was clear bringing a peaceful calm.

"Good evening, gentlemen," the agent said with a monotone, sterile voice. "What have we here?"

"I've got a fare for the Canadian side."

Looking through the driver's window the agent said, "May I see your papers, please."

"What papers?" Brian asked before he realized the agent was probably talking about his passport.

"Step out of the cab, please," the agent responded instantly.

Brian regretted his quick response but knew he had to remain calm. "Do you mean my passport? I have it right here."

"Please step out of the cab," the man said with a stronger, more commanding voice.

At least the rain had stopped. Brian opened the door and got out of the taxi. The air was cooler than he expected, but he was thankful for the extra stimulus to keep the fatigue away and concentrate his thinking.

The agent was about six inches shorter and considerably thinner than Brian. Cocky was the one word Brian would have chosen to describe this servant of the people. Bad feelings began to fill his thoughts as he visualized a wild set of scenarios of what was going to happen.

Holding out his passport, Brian stood by the open car door. The agent just looked at Brian for the longest time which made Brian feel very awkward and vulnerable. Was this customs agent going to arrest him on the spot? The struggle with his emotions was almost more than he could bear. Maybe he should give himself up to the authorities and go home. The yearning for the safety of his home began to overpower all the other emotions coursing through his consciousness. He wanted this tension to end. Nearing the limit of his tolerance, Brian was about to give up when the agent took his passport and began leafing through it.

Several times the customs man looked from his passport photograph of Brian. "You're under age, young man," the agent stated. "I can't let you out of the country without your parent's permission."

It's all over, Brian said to himself before he remembered the letter Malcolm had written for him. "I do have permission," he said as he quickly started to reenter the taxi to retrieve the letter.

"Stop. Don't move," the man commanded.

Brian froze in a bent over position half in and half out of the taxi. He looked back at the customs agent who now had his hand on the pistol holstered on his right hip.

"Don't ever move that quickly around law enforcement officers," he admonished Brian. A large flashlight came off the hook on his left hip. The dark overcast and the declining daylight made the interior of the taxi quite dark. Shining the hand light into the back seat the government man was reasonably assured there was nothing sinister, but he still wanted to proceed with caution. "Slowly get whatever you think gives you permission."

Brian's heart pounded as he found Malcolm's letter and produced it for the agent. The expression on his face as he handed the letter back to Brian was not one of thorough acceptance.

"You could have written this yourself."

"But, I didn't."

"How do I know?"

"Can you call my father?" asked Brian in a final, big bluff. However, he was prepared to give the man the number of the Bainbridge house.

"You're a long way from home. That would be an expensive call for the government," he said without looking at Brian. Instead, for some strange reason, he looked at the taxi driver sitting in the driver's seat patiently waiting for the conclusion of the interrogation. "Who are you going to see in Canada?"

He had rehearsed this routine numerous times with Malcolm in the last few days before he left Wichita. "I'm going to see some friends of my family."

"Like who?"

"Melba and Henry Jamestone."

"Where do they live?"

"In Montreal."

"Maybe we should call them."

"If you please."

The customs agent stared at Brian intently trying to sense the slightest possible indication of a falsehood. Malcolm had told him to believe, to know, and to look any interrogator directly in the eyes and to not look away.

"Fine," he said after what seemed forever to Brian. "Enjoy your visit with Mr. and Mrs. Jamison."

"It's Jamestone," Brian corrected the man as Malcolm had told him to do. "And, they are brother and sister, not husband and wife." Then, he wondered if he had offered information he was not asked as the taxi driver had warned him not to.

"Whatever," the agent said with a smile of some assurance the story was true. He motioned with his hand and said, "Get along now. I've got other folks waiting to see me."

As they started across the long bridge, Brian finally took in a deep, releasing breath that helped to slow his heart rate down.

"Now, that was one of the best performances I have ever witnessed," the driver said.

"My friend, Malcolm, helped me prepare."

"He did good and you did great, except for a couple of mistakes."

The water of the Detroit River was very dark as they crossed to the Canadian side. The lights of the two cities and the bridge added to the running lights of a ship passing beneath them.

"What mistakes?"

"Never move without asking or being told when you're around cops and correcting the agent's 'who-you're-visiting' trap was borderline."

"Well, at least, we got passed."

"Yep. Now, we're coming to the Canadian check point. They should ask to see your passport, your papers. Don't give him anything else unless he asks."

"OK."

The border station looked quite similar to the U.S. side except the men were wearing red coats with brown straps across their chests and peaked brown hats with a wide brim all the way around. They were also armed with pistols that were tied by white cords around their shoulders.

"Good evening," he said without much of an accent. "Passport, please."

Brian handed his passport through the window to the man. He quickly compared the photograph to the face in the back seat. Satisfied, the Canadian border agent stamped a page and handed it back to Brian.

"Enjoy your visit to our country."

"Thank you."

―

Sunday, 4.June.1939
Windsor, Ontario, Canada

"Here it is, Maple and Lewis Streets. So, what now, kid?"

Brian looked to the building on the northwest corner as Group Captain Spencer had told him. There it was, a two story, stone building. The entrance was on the South side in the center of building. The large white sign with black letters over the door read, EMERY OFFICE BUILDING. That had not been mentioned by John Spencer.

"I need to go into that building over there," Brian said pointing to the Emery Building.

There were no people anywhere in sight. Only one simple electric light with a wide conical reflector shield hung by wires over the center of the intersection provided the limited illumination. The driver parked directly in front of the door.

Brian tried the door. It was open. Without saying anything to the driver, Brian went inside. The hall was dimly lit with two lights made up to look like lanterns hung on the wall. The stairway was in the middle of the building just as Mister Spencer had said. The second floor was an almost exact copy of the first floor. It smelled fresh like a spring wind had blown through although a slight twinge of dust made it more real. There were four doors with frosted glass windows in them, each with a label painted on the glass. All the way at the end, right above the street entrance on the left, or east, side of the corridor was the door with the words, COMMONWEALTH EXCHANGE OFFICE, clearly labeled on it.

Brian returned to the waiting cab driver. "This is the place," Brian said proudly and with an edge of excitement in his voice. "How much do I owe you?"

"You can't stay here."

"I'll just stay by the door until they arrive tomorrow morning."

"Look, boy, haven't you got enough money for a hotel. I'll take you to the closest hotel."

"I don't have much money left. I don't need a hotel. I'll be all right."

"The hell you will."

After some discussion, Brian paid his fare which was about what Malcolm had told him to expect, $2.25. Brian sensed a concerned feeling in the man who had helped traverse the greatest obstacle. The man offered to buy Brian his dinner if they could talk. The excitement and suspense of the day's events began to take its toll on young Brian Drummond as the feeling of safety returned. The urge to just go to sleep rose so rapidly, Brian was not sure he would be able to stay awake through dinner. Sleeping was becoming more important than eating even though he was very hungry.

Nonetheless, Brian felt an obligation to the man. The taxi driver, whose name was actually, Jerry Rivers, had fought in France with the American Expeditionary Force under General Pershing, been wounded several times and returned with a serious limp. Driving a taxi had helped him since it was mostly sitting, but it had also contributed to the weight gain he was not particularly proud of. He accepted what life had dealt him and he was generally a happy man. Although Jerry Rivers did not look anything like Malcolm Bainbridge, the man reminded Brian of Malcolm.

Beside Brian's questions about combat in France, Jerry was intensely interested in flying and Brian's dream to join the RAF. Brian's enthusiasm for flying and the clarity of his dream managed to keep the conversation lively and to keep him awake for the hour and a half of their meal. The heavy eyelids and the occasional head jerk were blatant signs of the steamroller effects of fatigue.

Jerry Rivers recognized the signs and accepted the reality. He said he did not want to end this particular conversation with a young man he had just met and yet who seemed to be so much older than his years. The younger man's excitement was strangely contagious. However, Jerry paid for their dinners that were actually more than he had made transporting Brian to Windsor, but it was well worth it in his mind.

Stopping in front of the Emery Office Building, Jerry turned around in his seat. "Wake up, Brian."

Looking out the window and taking a moment to realize where he was, he said, "Thank you, Mister Rivers. You have been most kind."

"No problem, son. Good luck. You blast the Krauts right out of the sky. Think of Malcolm and me, and the others who fought the Germans just a few years ago. Win this one for us, too."

"I'll do my best."

"May God be with you and protect you, young Mister Brian Drummond," Jerry Rivers said with considerable emotion as he wiped away a solo tear descending his round and reddish cheek.

Brian waved and entered the building. He heard the taxi drive away and missed Jerry Rivers as if he was a good friend. Not wanting to miss the earliest opportunity, Brian had planned all along to doze outside the door to the Commonwealth Exchange Office.

Stretching out on the wood floor and resting his head on his small bag, Brian instantly felt the relief in his legs and back. He was also stone-cold unconscious before he had taken two deep breaths.

―

Monday, 5.June.1939
Windsor, Ontario, Canada

C<small>OLD</small> water on his face brought Brian back to consciousness as he spurted and sat up. Standing over him illuminated by the morning sunlight streaming through the window onto the far wall was an incredibly attractive young woman holding an empty glass. She did not appear to be much older than he was. In fact, at first, Brian thought she might be a school girl.

The woman's light brown hair was tied into a tight roll on the back of her head. The powder blue and white checks of her long calico dress made the soft white skin of her arms and face blend into the dress. Only the brilliant red of her full lips and the intense green of her eyes stood out.

"Why did you do that?" asked Brian finally.

"I could not wake you."

"Oh, come on. It couldn't have been that bad," protested Brian.

"I tried several times to shake you. I even kicked you a few times and if it was not for your color, I would have thought you were dead."

"Well, I'm sorry."

"Now, what can we do for you?" she asked with a soft, kindly voice that felt good to his ears.

A touch of curiosity grabbed him as he stood up and stretched his back. Brian was nearly a foot taller than the young woman. "Aren't you going to ask me to leave, or what I'm doing here?" A soft giggle was not what he expected.

"Oh, I think I know why you are here. This is not the first time we have found an American sleeping at our door."

Then, he was not the first, Brian told himself. Well, of course not, he continued. Why would they have an office, if they had no business?

"Do you want to come in?" she asked.

"Sure."

The office was not large. Two desks faced the center of the room on the far and left sides. The flag of Great Britain was on the left side of the far desk and a Canadian flag on the right. Above the desk between the two flags was a large photograph of King George VI. Three five-drawer file cabinets occupied the common corner and faced the other desk.

"Would you care to have a seat?" the woman asked motioning toward the chairs on either side of the window.

Brian dropped his bags to the right of the door and looked out the window onto Maple Street. People were moving in all directions on the streets below. Turning to the desk opposite the window, Brian saw the woman sitting and removing papers, pens and other office utensils from the drawers and placing them in specific places on the desk.

"You must be, Miss Helen Riser."

"I am. And, if I'm not mistaken, you must be Mister Brian Drummond."

"That I am."

"We have all of your paperwork here. Everything seems to be in order, and I presume you are here to volunteer for the Royal Air Force."

"Yes, I am."

"I'll need to get some information from you before Mister Blackwell can accept your oath."

Typing directly onto a series of forms, Miss Riser took down Brian's passport number, address of origin, vital statistics along with other bits of information about him. She was very efficient at transcribing the information.

"There we go," she said finally. "That should do it."

"Great."

"May I ask you a question?"

"Sure."

"Why does someone so young want to join the RAF?"

"I have a very good friend who flew for the Royal Flying Corps in the Great War."

"But, you are so young."

"I suppose so."

"You must be a very good pilot for an RAF group captain to take an interest."

"Some people seem to think so."

A short, slight, bald man about the size of Helen Riser walked into the office. He was neatly dressed in a medium gray suit with a white shirt and solid blue tie. He had a stiff salt and pepper mustache that was curled up on the ends into two small circles on either side of his nose.

The man looked first at Miss Riser, and then at Brian. "You must be Brian Drummond, I presume," he said with a similar accent as Group Captain Spencer.

Brian surmised that Reginald Blackwell was an Englishman and the attractive Helen Riser was a local Canadian helping with the administrative duties. Maybe he would ask her when the right moment arrived. However, there were other tasks immediately ahead. "Yes, I am."

"Excellent. Reginald Blackwell, here," he said extending his right hand. His grip was not very strong.

"Mister Blackwell," Helen said trying to get his attention. "I found him camped out by the door when I arrived this morning. I suspect he needs a clean-up and judging from his near comatose condition I found him in this morning, he probably needs a good rest."

"Quite right. We shall book you a hotel room as soon as we are done with the formalities. Is that all right?"

"Yes."

"Good, then, let's get to it, shall we."

Mister Blackwell read through a prepared briefing on the contract between a foreign volunteer and His Majesty's Government. All the conditions were just as John Spencer had said in St. Louis. Brian agreed and signed the form. Next, Brian was given an itinerary of steps he would take to get from Windsor, Canada to RAF Brize Norton, his initial flight training post. There was also a letter from John Spencer congratulating him on joining the Royal Air Force to assist in the defense of Great Britain in the approaching conflict.

"As luck would have it," said Blackwell, "you have a choice. We have a train leaving for Montreal this afternoon, or you may stay here, as our guest, until day after tomorrow so you might rest up from your arduous journey."

Although Brian had told neither Reginald nor Helen of his ordeal over the weekend, the fact that he was eighteen, alone and having traveled from Wichita, Kansas to Windsor, Ontario in two days was probably all the clues they needed. The thought of staying in Windsor for a few days and possibly getting to know Miss Riser was reasonably enticing. Brian considered the alternative, but discarded it quickly. He was only part of the way to completing his journey.

"I'd like to leave right away," Brian said. He thought, or maybe he imagined, he caught a slight, very brief expression of disappointment in Helen Riser's eyes.

"Very well, then."

With all the proper congratulatory statements, Brian accepted the oath to defend the United Kingdom in the service of His Majesty, King George the Sixth. As Reginald carried on some small talk about flying and the new Spitfire and Hurricane fighters, Miss Riser prepared a complete portfolio of the necessary papers, vouchers and other materials for the remainder of his trip.

After saying good-bye and thanking them both for their assistance, Helen Riser escorted him the four blocks up Maple Street to the Exeter Hotel. Without stopping at the desk, she led Brian up stairs two floors to Room 312 and opened the door with a key in her possession. The Commonwealth Exchange Office must be keeping this room for transients like himself.

Brian wondered if there was more to Helen's personal attention than professional courtesy, but did not have the courage to find out. She departed saying she would return in four hours to take him to the train station for the next leg of his passage to England.

The hot bath took all the stiffness out of his body and the resiliency out of his mind. The shadows of fatigue came back to him. He laid down for a short nap after setting the mechanical alarm clock on the table beside the bed.

—

I<small>N</small> the next instant of his awareness, a knock at the door was followed by the voice of Miss Helen Riser. "Mister Drummond, are you ready?"

"I must have fallen asleep, Miss Riser. I'll be ready in a jiffy."

"We need to leave soon. Your train departs in less than an hour."

"I'll hurry."

Helen waited in the lobby as he came down five minutes later. She had a car and drove several miles to the station. It was the first time he had ever seen a woman drive an automobile. There was something magnetic about it. Maybe it was the fact that he was so close to a beautiful woman while she had to watch the road and he could watch her. She did not seem to mind. Bumps in the road helped draw his attention to her ample bosom which giggled elegantly with each jolt. The gap in her blouse allowed an unavoidable glimpse of the lacy curves of her brassiere.

The railway station was on the north edge of town. The station was a small building with a broad roof overhanging a wooden platform next to the tracks and was quite similar to many small stations in the United States. Three rooms comprised the entire station house, a railway office for the ticketing agent, a lobby with bench seats for the passengers and a closed warehouse room for the baggage and freight of the railroad.

Most of the twenty minutes they spent waiting for the train were absorbed talking about flying and the prospects for peace in Europe. Brian wanted to know more about Helen, but she kept moving the conversation back to him. He was thoroughly enjoying this beautiful, sophisticated and intelligent woman, and especially one who seemed to be interested in him.

The whistle of the approaching locomotive brought a wave of regret to Brian. He wanted to capture the moment, but he knew it would only be a memory as a more intrusive calling beckoned him.

"Well, it's about time for you to go," Helen said as the train came to a stop. "They don't stop long here."

"I guess so."

"It was a pleasure meeting you, Brian. I hope you have a safe journey, and I wish you the best of luck."

"Thank you, Miss Riser. It was great meeting you." He extended his right hand.

Helen Riser must have felt more than just a professional duty. She also knew this young man was going off to a situation he was not likely to survive. She too believed the minority. Taking his hand, Helen Riser leaned forward raised up on the tips of her toes and kissed Brian on the cheek.

Her lips were warm and soft on his cheek, and yet the simple contact burned into him. He could smell the gentle scent of this woman, and it made his heart pound a little harder than it already was.

"Godspeed, Brian Drummond."

The young aviator from the Great Plains of America who thought he was in command of his environment was frozen solid like a block of granite chiseled into the shape of man. He felt foolish, but he could not do a thing about it. There was a part of him that did not want to move.

"You'd better go, or you'll miss your train."

"All aboard," shouted the conductor.

"Thank you," he said.

Before he knew it, Miss Helen Riser was fading away as the train pulled away from the station.

—

Chapter 3

> Beauty -
> the adjustment of all parts proportionately
> so that one cannot add or subtract or change
> without impairing the harmony of the whole.
> -- Leon Battista Albert

Friday, 16.June.1939
Southampton, Dorset, England

THE sight of land after ten long days at sea was marked by the steam whistle screaming into the air as it signaled the ship's arrival through the thin fog. The tension within Brian continued to grow with each moment of the passage across the Atlantic. The incessant motion kept him constantly uncomfortable, verging on nauseous, but never very sick. The confines of the freighter, RMS *Liverpool Lady*, although large by ship standards, so the crew said, on the vast expanse of ocean gave him a profound sense of his diminutive state. The majority of the strain came from the discussions.

The crew numbered nearly twenty, all citizens from the various corners of the British Empire. There were a couple of Indians, a Chinese steward and a South African while the remainder were from the British Isles as far as he knew. Only twelve passengers made this particular trip. The ship had accommodations for thirty crew and twenty-six passengers, so the first mate told Brian. He never did figure out where they would possibly berth all those people. The ship just did not seem to be that big.

During the endless hours of idle time, not more than an hour could pass without some talk of war. Surprisingly enough to Brian, the passengers were split down the middle expressing their opinions about the prospects for peace or war. A good number of the crew, Brian never did determine how many, were virtually unanimous in their belief war was coming.

Much to Brian's delight, most of the crew and passengers were complimentary and supportive of his purpose for this journey. With the exception of one man, Brian was treated as if he was somebody very important. Several people used the term royalty although he really had no idea what that meant. Especially for those who believed war was coming, the presence of a young, future fighter pilot who would help defend Britain from the same aerial onslaught unleashed on the hapless Spaniards just a few years earlier was reason enough for respect and recognition.

The one man who did not make Brian feel welcomed and appreciated was a Scotsman from Glasgow. He felt rather strongly the talk of war was the personal contrivance of Mister Churchill. He also felt Adolf Hitler was the best thing to happen to Germany and Europe in his lifetime. His wife clearly did not agree with him although she chose not to speak her disagreement.

All the attention, discussion and debate produced two distinct results. Anticipation was the dominant emotion. Brian wanted to get to the task of flying, of becoming the fighter pilot all these people were so impressed by. The other emotion was a strange mixture of anxiety, apprehension and a measure of fear from the unknown, the prospect of actual combat and the consequences of failure.

The land off the port side of the ship was the Southern coast of England as many of the people described it. Brian knew he was close. The thin, low overcast and haze obscured some of the details of the shoreline, but it was there nonetheless.

The Isle of Wight soon occupied the starboard horizon until the ship turned into the Solent estuary. Fishing boats leaving as well as returning to port moved gracefully around the several ferries and the *Liverpool Lady*. The chief cook, an Irishman, stood with Brian Drummond along the railing pointing out the topographical, nautical and historical features of Southampton and its approaches. Segments of the medieval port city wall were visible among the more modern buildings.

They would soon be docking in Southampton, the largest port on the South coast. The harbor was around them now and the buildings along with people walking on land provided its own excitement. Brian was glad this part of the journey was over.

The crew moved quickly and adroitly about their duties as the ship crept toward the pier. Brian could see the mass of people crowded on the dock waiting for their arrival. Some of the people were workers, stevedores as the crew called them, while the others were friends and relatives of the passengers. Brian hoped someone was there to meet him. He wanted to see John Spencer again, but he was an important man with great responsibility.

As the docking procedure progressed, Brian's thoughts wandered back among the twelve days of his odyssey from the plains of Kansas through the great cities of St. Louis, Chicago, Detroit and Montreal. He had been helped along the way from Windsor to Southampton by friendly caring people who made him feel good about what he was doing. They were appreciative. Fortunately for him, no U.S. law enforcement officers stopped him or even talked to him other than the kind Detroit policeman returning his penknife.

Brian was eager to get back into the air. He knew he had only a few days more to reach some airfield in Great Britain where he would start to fly again. The captain of the *Liverpool Lady* told Brian several times he thought war was imminent. It could start at any time which was why he was so edgy and on guard. If the captain was correct, Brian desperately wanted to be ready. In his own way, Brian hoped and prayed the war would wait for him to finish his training which John Spencer had said would take about nine to ten months, plus or minus a little, depending on how quickly he mastered the requisite skills and demonstrated his operational capabilities.

The docking process seemed to take forever, but it was finally over. Soon, Brian would set foot in England. His inner voice told him this was where he had to be, with memories of Kansas and his family far away at the moment. He stood by along the railing near the quarterdeck, the place where everyone would disembark. Brian had decided to wait for most of the people to leave before him, so things on the dock could sort themselves out making his task easier.

Searching the faces looking up at the passengers lining the railings yielded not a single familiar face. Brian knew he might not have someone waiting for him. The instructions provided by Reginald Blackwell in Windsor, Canada, told him someone would meet him at the dock, or he was to book a room at the Excelsior Hotel and call a specific telephone number. With those instructions, he wanted to allow as much time as possible for someone to meet him.

Most of the passengers disembarked into the waiting arms of friends and family. Several groups of dock workers had come on board probably to help prepare the ship for unloading. With a lull in the traffic across the brow, Brian sought the captain, thanked him for the ride and left the *Liverpool Lady* behind him as he descended to the wooden planks of the dock. The firmness of the pier felt strange beneath him as if he was still moving, swaying with invisible waves.

Several small groups of people stood around several of his fellow passengers. They were probably waiting for their baggage to be off-loaded. As Brian worked his way through the people on the dock, a familiar face emerged from an office. It was the bald head and round face of Group Captain Spencer in his RAF uniform this time. The most distinctive item on his entire uniform was the white crowned wings above the left breast pocket of his gabardine blue uniform. Brian remembered the photographs of Malcolm during the War. They were the wings of a pilot in the Royal Air Force.

"It is grand to see you again, Brian," John shouted across a few clumps of people. He was walking with a confident, purposeful stride as he had seen in St. Louis. John extended his right hand to Brian. As they shook hands,

John said, "Welcome to England, the land of your ancestors."

"Thank you, sir," Brian answered with some confusion over how this RAF officer might know his ancestry.

"My apologies for not being on the wharf for your arrival. I have, unfortunately, been on the telephone. Urgent Air Ministry business, they told me."

"That's all right."

"Well, are you ready to go, or would you care to freshen up."

"I'm just fine, sir. Actually, I want to get in the air as soon as I can."

"Ah, a good pilot to the heart. We shall fulfill your wish in a few days. First, there are a few formalities we must to tend to, I'm afraid."

The curious expression on Brian's face was all the confirmation John Spencer needed.

"It's near mid-day, we should have some lunch. Then, since we are so close, I would like to take you across the river to Woolston and the Supermarine factory."

Brian recognized the name and knew what it was associated with. The curious expression was instantly replaced with an enormous grin. "Spitfires."

"Yes, quite right you are. Spitfires. I should like to sit you in the cockpit of one of those beauties as well as show you around one of the factories, I believe, you will be defending in the not too distant future."

The mixture of the pleasures of aviation with the grim reality of the approaching conflict was awkward and disturbing to young Brian Drummond. The smudge of anxiety was also quite recognizable. The thought of touching the vaunted fighter he had only heard stories about soon took the principal focus of Brian's imagination.

John Spencer continued, "Shall we find something to eat?"

The question brought Brian back to the moment. He was not very hungry. "If you like."

"Very well, then. Shall we?" the RAF group captain said motioning toward the street at the far end of the dock.

They walked to a small corner pub not far from the dockyards. Talk of the coming events was liberally interspersed with their small meal of fried fish and potatoes Group Captain Spencer called, fish & chips.

After the midday meal, the plan called for a short taxi ride via the 'floating bridge' across the River Itchen to Woolston, visit the Vickers Supermarine aircraft factory, the birthplace of the Spitfire fighter, and then a train ride into London. The next day would involve another briefing and questioning session followed by the swearing in as a pilot candidate in the

Royal Air Force.

John Spencer told Brian several times his commissioning as an officer was quite out of the ordinary. Normally, Brian was told, he would be a non-commissioned officer, a flying sergeant, since he had not completed university. There were several reasons for the exception in Brian's case, but most notable was his flight experience. Brian sensed there was more to the exception than his flight experience. His relationship with Malcolm Bainbridge, the debt John Spencer felt for his American compatriot and Malcolm's service to the British Empire during the Great War had to be factors as well, Brian suspected. He also learned for the first time, Pilot Officer Malcolm Bainbridge had been awarded the Distinguished Flying Cross, twice, for bravery in the skies over France. For whatever reason, Brian Drummond was thankful for the consideration and the extra special treatment.

After a day in London, a couple of additional visits were planned prior to being posted to his first duty station, the RAF's elementary flight training airfield at Brize Norton in Oxfordshire. The events John talked about were a blur to Brian. He did not appreciate what was happening or imagine what was going to happen other than he was going to be flying soon. His thoughts were on the most immediate objective, seeing the Spitfire fighter for the first time.

"What do you say, let's go look at a real flying machine?" John finally brought the discussion back to Brian's level of attention.

"Great."

"Instead of taking a taxi, it's only a short walk to the ferry, the 'floating bridge,' landing. How about a walk?"

"Sounds fine to me."

—

Friday, 16.June.1939
Vickers-Supermarine Main Factory
Woolston, Dorset, England

THE Woolston area was more of an industrial area with larger warehouses and brick factory buildings with names like Southampton Marine Supply and Evergreen Pipe and Foundry. They stood before a high brick wall with one large wrought iron gate with smaller personnel gates on either side. The only sign for this area was an ENTRY PROHIBITED SEE WATCHMAN sign on either side of the three gates.

Group Captain John Spencer talked to the gate guards showing some type of identification. The security associated with this place told Brian it was important.

There were several introductions to men who had titles like general manager and assistant general manager. Then, Brian met the chief engineer, a mild-mannered man by the name of Joseph Smith, the same name as the Mormon leader from his American history books. The man did not care much for the title since he was not the real designer. Just from the words used to describe the actual chief designer of the Spitfire, Sir Reginald J. Mitchell, CBE, it was obvious they all had considerable respect for the man. According to Mister Smith, he had an unexpected and untimely passing in 1937, as the fruit of his genius was entering production.

Brian just wanted to see the airplane. Nearly an hour passed before he began to see the first indications they were in a factory that manufactured aircraft. First, it was the wing assembly area, then the landing gear. The machine shops grinding away metal, the sounds of rivets being bucked and the mechanical whirr and clank of the factory made the various buildings project a sensation of urgency. The Supermarine plant seemed to be almost identical to the Beech plant in Wichita. The layout was different, but the basic activities looked, sounded and smelled the same. The rich, deep aroma of lubricating oil pervaded every space just as it did at Beech.

Mixed among the mechanical sounds and the words of explanation from their guide, the assistant to the general manager, were greetings from the workers. Most stopped their tasks for a few moments to say as they passed, "We're buildin' 'em for you, mate," or "Use these for England," and many others. Brian was impressed and overwhelmed by the friendliness and sense of commitment he felt in the plant.

As they walked toward the large open door, the light brightened although the sky was still overcast. Stepping outside, the shape caught his attention immediately.

"My God, she's beautiful," Brian said with conviction of a man in love upon seeing the curves of his unadorned lover.

"Mister Drummond, this is the Vickers Supermarine Spitfire Mark One," the assistant said, as if he were introducing an important celebrity.

Brian could not remember the man's name and regretted not paying more attention to the introductions. He chose not to answer the introduction.

The aircraft was smaller than Brian had imagined. Painted with irregular, alternating patches of a dark brown, earthy color and a dark forest green with a blue, white, red and gold set of concentric circles, like a target bulls-eye, just behind the wing made the airplane look like a fighting machine.

Brian practically ran to the machine running his hand down the leading edge of the left wing. "Aluminum," said Brian half stating, half asking.

"Correct, sir. Formed aluminium with flush rivets to give her minimum drag and maximum speed," the assistant said.

"She's beautiful," Brian said with clear appreciation and totally oblivious to the broad smiles of the two older men. He stopped at the holes and protrusion in the leading edge of the wing.

"The business end of the machine, Brian," said Spencer. "She's got eight Browning, three oh three caliber, machine guns."

"My God."

"This is truth or consequences time, Brian. These guns mean deadly business, not like our flying in St. Louis."

"Yes, sir."

"Are you sure you want to get into this nasty business?"

"Yes, sir. I can feel it. I know this is what I was born to do."

The young pilot kept at least one hand on the sleek machine and sometimes with both hands embraced the Spitfire. He looked at the retractable landing gear – they called the undercarriage – the large coolant radiator under the right wing, the smaller oil radiator under the left wing and the engine air inlet under the centerline. An enormous, two bladed, fixed pitch, wooden, propeller dominated the nose. The smooth, aluminum skin made the gentle curves much more pronounced. The six large exhaust ports on each side of the forward fuselage were indicators of the powerful Merlin engine under the cowling, and they caught Brian's attention. Although he could not see the engine itself, he had never seen such a large exhaust manifold. This was without question the most powerful aircraft young Brian Drummond had ever been able to touch. It was also the best looking machine he had ever laid his hand upon. The fighter simply looked fast, very fast.

"We've got variable pitch props in the U.S.," Brian said more asking than stating as he touched the lower blade.

"There are some changes coming," responded the assistant with some defensiveness to his voice.

"A new prop?" asked Brian.

The man looked at John Spencer. "I can't answer that question."

"I can," said Group Captain Spencer. "Brian, this is sensitive information that I will give you as an RAF officer."

Brian nodded.

"We have numerous improvements in test for both the Spitfire and the Hurricane fighters. A three-bladed, variable pitch, constant speed airscrew along with a more powerful, eleven hundred horsepower, Merlin engine are among the changes planned for next year. We have a four-bladed airscrew in design, but the proper engine is not ready, and we need the airframes, now."

"I understand," responded Brian. His curiosity wanted to confirm the strange term John used. "Airscrews is that what you call the propeller?"

"Yes. You will find there are several differences in terminology here in England. You will need to get used to the changes."

"Airscrew. OK. No problem." Brian's examination was thorough and appreciative. "She's got cloth control surfaces."

"Quite right," their guide responded. "Designed for reduced hinge moments."

Brian did not exactly know what the term meant, so he continued around the tail with at least one hand still on the graceful fighter. Stopping behind the left wing, he looked up at the caged and glassed enclosure of the cockpit.

"Would you like to climb in her?" their guide asked.

"Sure!" exclaimed Brian with the characteristic excitement of his youth.

"I will open her up," John offered to save the extra effort for the overweight assistant.

John Spencer jumped up to the root of the left wing, turned the small handle, pulled the canopy back and lowered the access panel to the cockpit.

"With such a tight cockpit, we found we needed to provide our pilots with an easier entry," John said as he released the access panel. John moved slightly forward of the opening, motioned toward the open cockpit and said with the elegance of his class, "Your future workstation, sir."

Brian moved like a streak of lightning and slid into the confined space. The young pilot quickly scanned the cockpit. The dark surfaces, black instruments with white numbers and lettering in front of him, the throttle lever on the left and the undercarriage controls on the right were easily recognizable. The control stick between his legs had a peculiar circle on top of the column. He had never seen a stick quite like it.

As Brian grabbed the control stick first below the circle, then with either hand on each side of the circle, John explained. "The circle gives you equal control with either hand in case you are wounded. In addition, the g forces this fighter can generate in maneuvering flight sometimes require both arms. Other than the circle, we call the spade, the stick works just like you are accustomed to. You'll also notice the leather straps on the rudder pedals." Brian looked. "Again, an accommodation for a wounded pilot. With one boot in the strap loop, you can make corrections in either direction."

"Wow," was all Brian could say.

"What do you think?"

"This is fantastic. I can't wait to get this beast into the air."

"In due time, my boy. In due time."

Brian looked outside the cockpit, first at the wings, then over his shoulder. John Spencer knew exactly what Brian was doing. The young pilot's imagination was now in overdrive as he evaluated the field of view, what he would be able to see in flight, in aerial combat with this airplane. It was not as good as the old biplanes he was used to flying, but maybe the speed and maneuverability of the fighter would make up the difference.

Soon, he realized the airplane was sitting in a relatively large, open area of concrete surrounded by buildings on three sides and the river on the fourth. "How do you get the aircraft in the air?"

Two strong laughs were the initial responses to Brian's simple question. Then, the assistant responded, "We complete engine and control rigging here. Then, we demate the wing, transport the two major assemblies to a nearby aerodrome at Eastleigh, reassemble them, flight check the fighter and deliver the Spitfires to the squadrons."

Brian Drummond felt a bit foolish. Of course that is what they would have to do. "That makes sense."

John swept his arm around the area. "This used to be a launch ramp when Supermarine made float planes, the famous S.6 Schneider Trophy aeroplane – the predecessor of the Spitfire."

"Do you have any other questions?" asked the guide.

"Can I start her up so I can hear the engine?"

John Spencer jumped in. "I think not. We need to get you some training first, I should think."

"OK."

"Well, if that's it, I am afraid I must beg your forgiveness. I have a rather full schedule this afternoon," their guide said.

"Quite right. Thank you, Mister Granville. We appreciate your taking the time to show us your factory and product," John responded.

Brian remembered the man's name, Nathan Granville, and felt a little embarrassed about not paying more attention to the introductions. "Thank you, Mister Granville."

"You are quite welcome, young man. We hope you can use our aeroplanes well."

"I'll do my best."

"I am sure you will. Now, if you will permit me, I will lead you back to the front gate."

"We can find our way, Mister Granville. We'll let you get back to your vital work."

"No trouble," Nathan Granville said motioning toward the alleyway between two buildings leading in the general direction of the main gate.

As they walked out, Brian looked curiously through windows and doors to see the activity within. He recognized more parts being built, now that he had seen the finished product. Brian looked over his shoulder several times as the impressive fighter disappeared behind the intervening buildings.

"Ever since I first saw this aeroplane several years back, I've thought its gentle, graceful, natural curves remind me so clearly of the elegant curves of a woman. Don't you think?"

Brian felt a brief flush of embarrassment, but soon grinned in recognition. "That's it, isn't it? Every curve does seem so . . . I suppose . . . natural." Brian liked the statement, not because it was so descriptive, but because it was man-to-man kind of talk, pilot-to-pilot. In a strange way, it made him feel closer to his goal.

After the final expressions of gratitude, John and Brian returned to the 'floating bridge' and Southampton. Retrieving Brian's bags, John ordered a taxi to take them to the train station. The Southern Railway journey through the country side of southern England was splendid with all the greens of late spring, the deep browns of freshly plowed fields and the small clusters of houses John called, hamlets. The gentle rolling hills reminded Brian of Kansas, east of the Flint Hills. In the idle periods between descriptions and conversation with John Spencer, Brian's thoughts returned to the people he left behind in Kansas, his parents, Becky, Malcolm and his friends. He was a long way from home and he missed the people he had grown up with. It was the idle time that brought regrets of his determined action. As soon as he was settled in one place for a reasonable amount of time, he would write to his parents and Becky to let them know he was OK.

During a lull in their conversation, John Spencer said, "I almost forgot . . . ," as he retrieved a yellow envelope from inside his uniform jacket, "this arrived for you, several days ago."

Brian opened the envelope.

TELEGRAM

UUUU/1248956/BTU-VGHU/23RTOP498ZZ/UUUU
7 JUNE 1939
TO: MR. BRIAN DRUMMOND
C/O: GROUP CAPTAIN JOHN H.R. SPENCER, DFC,

```
            BUSHEY HEATH, MIDDLESEX, ENGLAND
            STOP
            TENSION REMAINS HIGH FOLLOWING YOUR DEPARTURE
            STOP I AM ACCUSED OF SUBVERTING YOU STOP
            HANDLING SITUATION OK FOR NOW STOP YOUR
            PARENTS NEED A LETTER FROM YOU SOONEST TO
            KNOW YOU ARE OK STOP BECKY ANGRY STOP SHE
            NEEDS LETTER AS WELL STOP MY BEST TO JOHN AND
            MARY STOP TAKE CARE OF YOURSELF AND ALWAYS
            CHECK SIX STOP
            SIGNED MALCOLM
            END
            UUUU/1248956/BTU-VGHU/23RTOP498ZZ/UUUU
```

Brian's pensive state triggered John Spencer. "Is everything all right?"

Brian took some time to answer. "Yes, I suppose so. Malcolm sends his regards to you and your wife." John nodded although Brian barely noticed. "I imagine my parents are making life very hard for Malcolm and for that I am deeply sorry."

"I don't think he would want that, Brian. He's a strong man, and he certainly knew what he was doing in helping you. He shares your vision, as do I."

"Yes, but I didn't want him to have any trouble because of me."

"Then, if you would permit me, maybe you should write to your parents on his behalf."

"That's what the telegram says." Brian thought about what he needed to say. "Well, not on his behalf, but to tell my parents I'm OK."

The more continuous number of houses, shops and other buildings signaled their entry to the outskirts of London. The train ride from Southampton, including the stops at cities like Eastleigh, Winchester, Basingstoke and Farnborough, took a little over two hours. Brian surmised he would soon be making the same journey in about ten minutes with his Spitfire flat out traveling at more than 400 miles per hour.

Disembarking from the train at Waterloo Station in the heart of London, they took a taxi across the River Thames to the Savoy Hotel along the Strand near the theater district. John made all the arrangements for his room.

"If you feel up to it, I would like to leave you for a few hours. Are you amenable to company for dinner?" asked John.

"Sure."

John smiled not quite accustomed to the peculiar choice of words the young American selected. "If you have no objections, I shall fetch Missus Spencer."

"That would be great."

"Very well, then, I shall be off," John said with a cheery voice. "See you in the lobby at half past eight."

―

Friday, 16.June.1939
Westminster, London, England

THERE was always a certain sensation of rejuvenation associated with a good bath, fresh clothes and the overall clean feeling. Stepping out onto the busy street brought a slight regression with the odors of automobile and bus exhausts attacking the nostrils. The light spring breeze did help to lessen the detractors and brought other smells – sweet edible smells.

Brian Drummond had half an hour before he was to meet Group Captain and Missus Spencer for dinner. Why not explore a little to get a feel of the city, he asked himself? His instincts were somewhat in overload with the hustle and bustle of the famous capital city. Brian remembered his history. London was the seat of the British government, Parliament, the home of the monarchy, the capital of England, of Great Britain, of the United Kingdom, and for that matter, the entire British Empire.

A short walk to the west with the setting sun's last rays upon his face brought Brian into a large open area like a piazza in Italy he had seen pictures of in his history book. An enormous column stood in the center of the square surrounded by large buildings. People moved in every direction while some did not move at all. Brian read the inscription at the base of the column. It was a tribute to Admiral Lord Nelson. The light of recognition burst on for Brian. This was Trafalgar Square, THE Trafalgar Square, and this was the monument to the most famous British naval hero, next to maybe Sir Francis Drake.

Brian was impressed, maybe even overwhelmed by the history, the majesty, the grandeur of this country. So many of the traditions as well as some very important events in his own country's history were entwined with England. For some reason, Brian's thoughts drifted back to a map he saw in school. All the countries that were a part of the British Empire were colored in red. He remembered a phrase his teacher had told the class. The words were written in the early 1800's by an English journalist whose name he could not remember. "The sun never sets upon the Union Jack," Brian said aloud to himself as he looked up to Lord Nelson atop the grand column.

"Quite right, old boy," answered a man dressed in a dark gray suit with a round black hat on his head.

Brian felt a little embarrassed and silly for acting like a school kid. But, then again, he had just finished high school, hadn't he?

The shadows of Trafalgar Square were deepening. Brian's thoughts were jerked back to the present with his recognition of the passage of time.

Another man dressed similarly to the earlier respondent was walking by. "Do you have the time, sir?" Brian asked.

"Not quite half past eight, I should think," he said as he pulled a large watch from his waist coat pocket. "Yes, quite so."

"Thank you."

"Yank, eh?" the man asked as he passed Brian.

"Yes, sir."

"Welcome to England," he said over his shoulder as he kept walking.

With only a few minutes, at best, to get back to the hotel, Brian half ran, half walked very fast, slowing down to a brisk walk pace for the last hundred yards. A series of deep breaths helped to slow down his respiratory rate and eventually his heart rate.

Among the throng of people moving up and down the walkway, the distinctive blue uniform of a RAF officer got out of an expensive looking automobile. Brian did not recognize the make of the car, but it looked like a limousine. The officer entering the hotel was John Spencer. Brian entered the lobby several paces behind him.

"Group Captain Spencer," Brian called out trying not to be too loud.

"Ah, there you are. Are you ready?"

"Yes, sir. I suppose I am. Am I dressed OK?" Brian asked conscious of dress codes in gentlemanly circles. He had put on his last clean shirt with his only tie and jacket.

"Yes, quite all right, Brian."

"Then, I am ready."

"Shall we go, then?"

"Yes, sir."

John led the way back out to the street and the waiting car. A man in a black suit and tie with a white shirt and soft, brimmed black cap was standing by the rear door. As he saw John approach, the man opened the door.

"After you, Brian," John said motioning toward the open car door.

Inside the dark interior was an immensely attractive woman older than Brian, but not as old as John Spencer in appearance. Brian sat down in the seat across the car from the woman who had to be John's wife. She was wearing a dark, probably blue, woman's suit with a frilly, light colored blouse. Only the lower part of legs were uncovered. A dark hat covered her hair which was drawn up under the hat. A broad smile was clearly highlighted by her full, red lips.

"And, you must be the Brian Drummond. John has told me so much about," she said in a smooth, elegant English woman's voice.

"Yes, ma'am."

"I am, Mary Spencer," she said extending her hand to him. "Welcome to London."

"Thank you," responded Brian taking her hand which was warm, soft and delicate.

As John settled into the luxurious leather seat next to his wife, Brian looked away from Mary Spencer. With the door closed and the chauffeur returned to the driver's seat, the car began to move into traffic and Brian was able to notice the interior for the first time. Soft, light colored, pleated leather seats faced each other. The passing lights of the city gave fleeting glimpses of the dark grained, polished wood panels around them. The interior smelled of leather, fine furniture polish, and lilac that must be Mary Spencer's perfume.

"Since you have been away from home for nearly two weeks," Mary spoke with such a delightfully melodic, soft voice, "we thought you might like a taste of home. There is a little cafe off the other side of Covent Garden that specializes in American food."

"That would be nice."

"John tells me you just turned eighteen a few months ago."

"Yes, ma'am."

"Why are you giving up your youth for this silly adventure?" she asked with serious concern.

John Spencer's eyes were on Brian, but he chose not to participate in this topic of conversation.

"Well, ma'am, I guess I don't feel like I'm giving up my youth."

"That may be, however in England you are not considered an adult, quite yet."

Brian began to feel uncomfortable with her words and the strange bestial attraction he felt for this woman. "That may be," he responded somewhat mocking her words, "but, I am a pilot, a good pilot and Group Captain Spencer," nodding his head toward her husband, "tells me you need good pilots."

"Yes, I suppose he would. Both he and his bellicose uncle are convinced Mister Hitler is on the warpath, as you say in your old West."

"I don't know much about politics. I'm just a pilot."

"Well, then, I should think we shall change the topic. How do you like England, so far?"

A peculiar set of thoughts rumbled through Brian Drummond's brain. Was Missus Spencer gently trying to be antagonistic, or was she simply baiting him to see how he would respond? Did John Spencer ask her to do this to see if he was really serious about flying in the RAF? Why was John subjecting him to this type of questioning? Brian suddenly did not feel like talking, but there was a question before him from his hostess. "It's nice," was all he felt like saying.

"Just nice?"

"Yes, ma'am. Sometimes it reminds me of parts of Kansas."

"Kansas, ah yes, America's heartland."

"Yes, ma'am."

"Excellent," interjected John Spencer. "Here we are."

The car stopped. They waited for the chauffeur to open the curb side door. John got out first, turned to offer a hand to his wife. Brian watched her move across the car and step out. Her skirt rose slightly above her knees revealing her firm and shapely legs. Brian followed Mary Spencer out of the car.

The narrow street was crowded on both sides by buildings rising two and three stories. There seemed to be shops on the first floors with apartments or offices above them. The lights and colors of the night mixed with the myriad of people moving along the street. There was no clear separation between the spaces for humans and for vehicles.

John held his hand out toward a narrow alleyway. The three walked down the alley with Mary in the middle. The Spencers took turns describing the area of London they were in along with some of its colorful history. There were small shops sporadically placed along the narrow street. At the far end, Brian could see a large open area, like an interior square.

"Here we are," John said stopping about two thirds of the way down the alley. "Digby's."

The colorful, red, white and blue, neon sign over the large, open, double doors with a bank of ground-to-ceiling, blocked windows just beyond marked their dining locale.

"Group Captain and Missus Spencer," the *maître d'hôtel* said loudly and with obvious enthusiasm. "A pleasure to see you, again. And, who is the young man, your son?"

"Good to see you, George," Mary said with her soft, sensual voice.

"No," John responded. "A friend and soon to be officer in the air force."

"Yes, yes, but much too young for war."

"But, an exceptional pilot," John said.

"Another flyboy, how nice."

"Brian, this is George Stafford. He owns this establishment. George, may I introduce, Brian Drummond."

"So good to meet you, Brian."

"A pleasure to meet you, Mister Stafford."

"No, no, just George," he said, and then looked to John with an exaggerated expression of surprise across his entire face. "Do we have an Yank?" George asked looking back to Brian.

"Yes, sir."

"No, no, please, please, just George."

"OK, George."

"Where are you from?"

"Kansas."

"Ah yes, wheat and beef. Am I right?"

"Yes, and aviation."

"Aviation, of course." George turned back to John. "I have your table ready, sir."

Walking through the modest size cafe, Brian was amazed by the number of military uniforms. The sounds were loud with laughter plentifully intermingled among the isolated words and fragments of sentences filling the room. The cafe was full, actually more than full, with numerous people standing in the corner around a bar. There was one open table by the front window.

"Your table, sir," George announced pulling one of four chairs back to seat Mary Spencer. With the three newest patrons seated, he said, "Enjoy your meal with us."

Digby's cuisine was up to the image John had presented. Brian chose a sentimental favorite, a Kansas City Strip, while John and Mary each had T-bone steaks. The meal was served with a baked, or as the English said, a jacketed potato and kernels of corn. The conversation was more in the mood of Digby's than the words passing in the limousine. There was no talk of politics, or war, or even flying, for that matter. The topics spanned the range from English beers and bitters to farming in general.

Brian was able to ask numerous questions about and learn of the privileged life of the Spencers, and the Armstrongs, Mary's family. No one had told Brian, John's great aunt, Winston Churchill's mother, was an American, and a rather notorious one from what little they said, named, Jennie Jerome.

She was popularly known as Lady Randolph Churchill, or just Jennie.

The meal was excellent and was the longest Brian had ever experienced, lastly about two hours. The conversation was light and jovial. However, Brian still detected a sliver of resistance, of discord, in Mary's tone or choice of words. The feeling in Brian was not conclusive and the young man had to accept the possibility he was somewhat biased by the earlier words.

With the proper congratulations and gratitude to the owner and the chef, the evening threesome departed Digby's turning away from the street and toward the square. It was a large square filled with people having fun, or just being together. A smaller set of buildings occupied the center of the square.

"This is the Covent Garden market square," John said waving his hand toward the square. "Some say, this is the most colorful place in all of Great Britain. Only in the dead of winter is there a break in the entertainment," he added referring to the impromptu musicians performing in different areas and the strange attire of some people, both young and old. The lighthearted talk matched the mood of the square. The rich aromas of everything, the activities, the people, the places, in the square were considerably less than serious. Brian found it was difficult to not fall victim to the tempo and excitement of Covent Garden. Nonetheless, he was thankful when Mary finally suggested it was probably time to leave. Brian was having fun, but he was also beginning to feel the fatigue of the endless onslaught of his senses.

Laughter was the dominant presence during the return to the car and the hotel. Brian acknowledged to himself that he felt much better about Mary and her attitude than he did at their introduction. As they arrived at the Savoy, John gave Brian a short set of instructions.

"You have a full day tomorrow, Brian. I'll pick you up at nine o'clock. We'll get you sworn in at the Ministry, and then I will turn you over to my assistant to get you outfitted with the proper uniforms. You pick up your flying kit at your first posting. Any questions?"

"No, sir. I've got it."

"Good," John said. "See you in the morning, then."

"Yes, sir." The door opened for Brian. "Nice to meet you, Missus Spencer, and thank you both for a great evening."

"You are quite welcome, Brian," responded Mary. "I look forward to seeing you again. Sleep well."

"Thank you," Brian said. He left the car and watched it drive away. The future RAF pilot was asleep in his room virtually within minutes.

—

THE drive to the Spencer's town house lasted a little longer and involved more discussion. It was amicable between two conspirators.

"Well, dearest," John began with an edge of sarcasm, "you certainly did as I asked, although I thought you might have come on a little strong with your *tête-à-tête*."

"Just trying to do your bidding, dear."

"What did you think of him?"

"A delightful young man."

"He is an even better pilot, actually one of the best natural pilots I have ever known."

"Really."

"He has the instincts we need for the fighter pilots who will soon be our only defense against the Nazi war machine. He's a natural hunter."

"He is so young, John."

"Yes, he is, but he is not the youngest. We have a few flying sergeants who are barely fifteen."

"They are just babies," she protested.

"Unfortunately," John responded as they passed St. James Park and Buckingham Palace. Group Captain John Henry Spencer knew better than all except a few people in the Air Ministry and Uncle Winston's inner circle of friends and confidantes, the severity of the looming threat. Winston had a few disciples and members of Parliament who shared his tabloid labeled, macabre view of coming events. John Spencer knew the survival of Great Britain as the world knew her would soon be in the hands a very few dauntless fighter pilots. The Home Guard was virtually non-existent and the prime of the British Army was deployed in Eastern France as the British Expeditionary Force, and the Advanced Air Strike Force. The surgical demonstration of aerial prowess by the *Luftwaffe* in the Spanish Civil War made the threat so much more real. "We need more pilots like Brian," John added with a combination of frustration and dejection.

"Well, then, I hope he flies the bloody hell out of those new aeroplanes you are building."

"He will."

Chapter 4

Duty cannot exist without faith.
-- Benjamin Disraeli

Saturday, 17.June.1939
Savoy Hotel
Strand
Westminster, London, England

"Good morning, Brian."

"Good morning, sir."

"I trust you were able to sleep well."

"Yes, sir."

"Excellent," John said standing in the modest elegance of the Savoy Hotel lobby. "We have a full day ahead."

The expectation was high. Today was the next day in the achievement of Brian's dream. Although the paper he signed in Windsor, Canada, said he was a member of the Royal Air Force, he did not feel the part. As John Spencer had indicated the previous day, Brian would be formally sworn in as a pilot candidate. Brian also expected to be wearing the uniform of his position at the end of the day. Then, he would feel the part.

As they exited the hotel, the doorman stood beside a rather plain, medium size automobile with the Regent's crown symbol and small white letters spelling, AIR FORCE, underneath on the front door. The doorman opened the rear door. John entered first.

"Have a wonderful day, gentlemen," the doorman said as he closed the door behind Brian.

"Brian, this is Sergeant James MacDougall," John said. "Sergeant MacDougall, this is soon to be Pilot Candidate Brian Drummond."

"Pleased to meet you, sir," Sergeant MacDougall said.

"Pleased to meet you, Sergeant."

With the introductions complete, the sergeant drove away. The journey was rather short. John reminded Brian of the day's agenda. Sergeant MacDougall would be Brian's escort and guide for the day's activities. They stopped in front of a small door to a large, cold, gray, stone building, like a side entrance with no signs around it.

"Mister Drummond will need you back here in forty-five minutes, sergeant."

"Very well, sir."

Group Captain Spencer exited the car and entered the building quickly causing Brian to practically run to catch up with him after closing the car door. Two armed soldiers occupied a very small lobby. The only objects to distinguish the place were a simple desk and a large photograph of King George VI, the same photograph he saw in Canada.

After the loud crack of their boots and a crisp salute, the desk sergeant checked John's identification and asked for him to sign in Brian since he had no official, Air Ministry identification and he was not in uniform. The security measures were quickly dispensed with.

Inside the heavy, locked, armor door, the hallway was larger than the ante-room and quite dark from insufficient lighting and dark, wood paneling. They passed down several halls, climbed two sets of stairs and finally entered an office marked, OFFICE OF THE SECRETARY OF STATE FOR AIR. Underneath the large permanent sign was a smaller wooden panel with gold lettering -- The Right Honorable Sir Kingsley Wood, MP.

A receptionist greeted them saying the minister was expecting them. Brian followed the RAF officer into a large, well appointed, homey office. A bald, round faced, bespectacled, older gentleman sitting behind the large desk concentrating on the papers before him, rose to greet them. As the older man stood to walk around his desk, Brian noticed immediately he was at least a foot shorter than he was.

"So, this is our fabled young American," the man said.

"Yes, sir," John Spencer responded.

"Minister, this is Pilot Candidate Brian Drummond," announced John, then added almost parenthetically, "not yet in uniform."

"Brian, this is the Secretary of State for Air, Sir Kingsley Wood."

"A pleasure to have you with us, young man," the minister said with a strong, energetic voice.

"It is an honor to meet you, sir," Brian said trying to overcome the tedium and burden of meeting so many people in such a short time.

Minister Wood asked essentially the same questions Brian had become all too familiar with over the last two weeks. He tried to be interested in the conversation and he wanted to answer with the same enthusiasm as the minister was exhibiting. Five minutes of general discussion seemed like an hour to Brian although he knew this was a genuine honor to meet someone so important. The older man talked about flying, and the interest His Majesty's Government had in building the RAF and specifically Fighter Command. The minister did not sound as convinced of the impending danger as Group Captain Spencer and others, but he was still saying the right words.

"John speaks highly of your skills as a pilot," Sir Kingsley said. "I truly hope we shall not need your skills although I suspect we may."

Brian did not know quite what to say. He finally fumbled for the only words he felt might be appropriate. "Thank you, sir."

"Quite so. Now, I know you would rather by flying than talking to an old stick in the mud. So, shall we get on with our duties."

"Yes, sir," responded John Spencer. "Shall I call Missus Treblewind to be our witness?"

"No need," he answered lifting the telephone. "If we may have the pleasure of your company, Missus Treblewind."

The matronly secretary entered the room promptly.

"If the situation on the continent continues to deteriorate, I may soon not be afforded the luxury of welcoming young foreign pilots such as yourself. Brian, you are the first American to join our little band of merry men. I suspect we shall see others, before long."

The minister with the two witnesses administered the oath of allegiance to the Crown. The formalities were dispensed with quickly. Brian felt a bit odd, having grown up in America reciting the Pledge of Allegiance to the United States of America every morning in his school, now having sworn to defend the King of England, the successor to King George III, whom his ancestors had successfully rebelled against. He rationalized the uneasiness as the price he had to pay for the opportunity to fly fighters, and some thought the most advanced fighters in the world, in combat. The image of the elegant Spitfire on the ramp at Supermarine in Woolston came to his thoughts. Mixed among the selfish thoughts of flight were the growing strength of feelings toward defending freedom against the forces of evil at the very front edge of that battle.

"Congratulations, young man."

"Thank you, sir."

"Just remember, all we ask is, you do your best. The good Lord shall take care of the rest."

"I will, sir."

"Brilliant, then we shall win the day when the conflagration comes. My best to your uncle, John."

"I will pass your greetings, Sir Kingsley, and thank you for your time."

With congratulations and good-byes said, John and Brian left the minister to his administrative duties. John introduced Brian to several more RAF officers before stopping at the personnel assignment office to pick up Brian's first orders – his posting to the Elementary Flight Training Unit at RAF Brize Norton for primary, or entry level, flight training. He was to report tomorrow.

John Spencer helped Brian find his way out of the building and into the waiting car driven by Sergeant MacDougall. They would meet again at half past eight for dinner with Mary. Getting fitted for his uniforms would take a couple of hours. Sergeant MacDougall was then tasked with a brief tour of central London before dropping Brian off at his hotel. The rest of the afternoon would be open for Brian to do as he wished until evening meal.

The assignments for the day were completed without a hitch. The tailors were impressive to young Brian. Everything was handled in a swift and professional manner with humor. The two tailors to the officers of the RAF made the experience actually enjoyable. The bill for the uniforms was added to a RAF account, once authorized by Group Captain Spencer and confirmed by telephone. Brian would have one complete uniform with a full complement of shirts, shoes and the other accouterments of officers in service of the Crown. The remaining uniforms would be delivered in a few days. Looking at himself in the mirror as the tailors worked their magic, Brian, for the first time, began to feel like a member of the air force although he did not have the distinctive wings over his left breast pocket, yet.

Sergeant MacDougall took a different route back into the center of the city. London was a large and beautiful city, maybe the most beautiful city he had seen, thought Brian. The tour was actually very well done as if James MacDougall was a professional tour guide. It was really a traveling car tour. Famous sights like the north clock tower affectionately known as Big Ben; the Palace of Westminster, otherwise known as Parliament; Buckingham Palace; the Tower of London and so many others were pointed out along with a little of the history and significance of the places they saw. James MacDougall was an outgoing, interesting person, older than Brian, but appropriately respectful. Brian was not quite accustomed to being addressed as, sir or Mister Drummond. It was nice for his impressionable, absorbent mind. Brian liked James MacDougall.

"Would you like to see anything else, sir?" Sergeant MacDougall asked.

"I don't know what there is," Brian answered. "You have shown me everything I was aware of in London, plus much more. This is such a great city."

"Well, Mister Drummond, there is a lot more to see in London."

"I've taken enough of your time. Why don't we call it a day?"

"I'm at your service, sir."

"In that case," Brian paused wanting to add a little fun to the parting. "Nah, just kidding, sergeant. Please drop me off at the hotel."

"As you wish, sir."

After stopping in front of the hotel, MacDougall stood on the sidewalk before Brian could get out of the car with his box of uniform items. "Thanks, James," Brian said.

"You are most welcome, sir."

"You have been really great, James. Good luck."

"You are not getting rid of me that easily, sir. I shall pick you up in the morning at oh eight hundred hours and take you to Paddington Station to catch a Great Western Railway coach to Oxford."

Brian immediately imagined an old West stagecoach like the examples he saw at the Cowtown Museum in Wichita, Kansas, or on the movie screen. "Great. Then, have a good evening and I'll see you in the morning."

"Cheerio, sir," MacDougall said as he saluted Brian and left.

After depositing the box in his room, Brian decided to take a walk down to the River Thames. It was not far away judging from the trees he could see down one of the narrow cross streets. The trees, grass and shrubbery brought serenity to what was otherwise a busy thoroughfare. Barges, tour boats and small ships moved up and down the river with clear purpose. People moved up and down the spacious walk along the bank. Some of the people were businessmen obviously enroute between meetings or appointments. Others were couples enjoying the warm sun and light breeze of the late spring afternoon.

Brian found a free park bench to sit, relax and watch the world pass by him. It did not take long before a torrent of thoughts began filling his consciousness. The thoughts were good thoughts, but they brought bad feelings. He missed Becky and Wichita more than he realized. In fact, he was quickly becoming aware of the coalescing reality he missed Becky more than his parents, Malcolm or his friends. Despite the excitement of his journey and England, and the attention he was receiving, questions began to roll through the flood of thoughts.

Maybe he had made a mistake? Why did he ever think answers could be found in England? Why had he done this? Maybe he should just admit this whole adventure was a mistake, so he could go home to the softness of Becky's touch? Brian missed Becky more than he thought he ever would. He could have flown in Kansas, flown for Malcolm and been happy. Why did he do this?

With his elbows on his knees, Brian buried his face in his hands. He felt the urge to let the tears within him roll down his face, but an unknown force would not let him release the emotions.

"I can't quit, yet," Brian said aloud to himself as he raised his head only to see an elderly man walk by directly in front of him. The flush of embarrassment filled his body.

"Certainly not," the old man said.

"I was talking to myself."

"What are you thinkin' 'bout quittin', laddie?"

"I joined the RAF."

"Aren't you a mite young for the air force?"

"I'm a pilot, and they need pilots," Brian said like a broken record he felt he had become.

"Is that so?" he said more rhetorically than inquisitively. "If my guess is correct, I'd say you were a Yank, or a Canuck maybe."

"American."

"So, you're here to defend Mother England, are ya now?"

"Yes, sir," Brian answered with some of the enthusiasm returning to him.

"Well, glad to have ya with us, laddie. Keep your spirits up and you'll be quite all right."

"Thank you, sir."

"Good day to ya, now."

"Good day to you, sir."

The old man walked on, leaving as quickly as he had appeared. The old man's words were just enough to bring Brian back at least for the moment. Brian walked up river a little finding an occasional rock to toss into the river. He began to think of the good things. London was a beautiful city. England and the British people were delightful. Then, the glorious image of the magnificent Spitfire sitting on the ramp at the Woolston factory finally brought a big smile to Brian's face and a quickening to his step.

The deep, robust gong of Big Ben signaled 18:00, enough time to take a quick bath and get ready for his dinner engagement with Mary and John Spencer. The thought of seeing the elegant features of Mary's face and straining to imagine the shape of her body made the intervening time pass quickly. Brian was in the lobby thirty minutes early to not miss a moment of time.

"Ah, there you are, Pilot Candidate Drummond," came the greeting of Group Captain Spencer.

"Yes, sir."

"Shall we be off, then?"

"Yes, sir."

As they walked out the front entrance to the Savoy, John said, "I'm afraid it's just us tonight, Brian."

Disappointment shot through young Brian Drummond. He even felt the tremors of guilt as if his earlier near carnal thoughts were the source of his current punishment. "No problem," Brian said with as much enthusiasm as he could muster.

The dinner was a new experience for Brian. John had chosen a suburban, corner restaurant, he called a pub, by the name of Regal Arms. He was also introduced to Shepherd's Pie and English bitter, a warm, dark beer. The conversation, aside from the explanation of the importance of the pub to the English, made the meal a working dinner.

John carefully explained what was going to happen over the next few days and weeks. The expectations would be high because of his citizenship and his connections. John told him repeatedly to never make, or offer, any excuses for either. Brian was told about the curriculum of his training. What things to watch out for, and when to be aggressive and when not. He covered as many of the customs, procedures and processes as he could remember. The most important advice was an echoing of Malcolm. Let your actions do your talking for you, John had said repeatedly. Questions and answers passed both directions well into the night.

By the time they left the Regal Arms pub, Brian felt light headed and a bit unsteady. It was the most alcohol he had consumed in one evening in his entire life. The time spent with John Spencer was well worth whatever punishment he had done to his body and brain. Brian liked John Spencer from the first moment they met in Saint Louis. Tonight, it was more like brothers, an older brother. It was a good time with an older brother, or at least what he imagined what buddy time with an older brother would be like. He liked John as he liked Malcolm...they felt almost the same.

"Well, Brian, tomorrow is the day," John said in the lobby of the Savoy.

"I can't wait."

"There are days I wish I was in your shoes, but we haven't figured that trick out, yet."

Brian had no way to actually appreciate, or understand what John had just said. He chose to say nothing and simply nervously scanned the room taking note of the characters.

"I will not be able to see you off tomorrow. So, I shall wish you good luck tonight. Our prayers shall be with you and we'll celebrate upon your completion of training and posting to your first squadron."

"That would be great, sir."

"So it shall be, then," John said. "Now, I shall bid you *adieu*. Give 'em hell, Brian."

"I'll do my best, sir."

"I know you will."

Sunday, 18.June.1939
Oxford, Oxfordshire, England

THE journey from London had taken the better part of the day. Sergeant MacDougall's staff car, two trains and now a small truck, the British called a lorry, were the vehicles of transportation for Brian. His destination for the day's travel was the RAF pilot training base situated conveniently in the English Midlands.

The Great Western Railway ride took several hours stopping at almost every station. The time offered Brian plenty of opportunity to consider his leisurely Sunday looking across the River Thames at the most famous clock in the world atop the clock tower of Westminster; the intricacies of Westminster Palace, Parliament building; the Imperial War Museum; the Tower Bridge and the infamous Tower of London. The 13.5 ton Big Ben bell produced the most regal time marker he had ever heard. Brian walked several miles, enjoyed every minute and was thoroughly exhausted at the end of the day. Sergeant MacDougall had provided very good guidance as well as perfect directions on a complicated map in a large city where there seemed to be no straight streets. London was truly a great city.

Despite the enjoyment and entertainment, he missed his family, friends and especially Becky. He removed Malcolm's telegram from his interior tunic pocket. He had to find some time as soon as possible to write to the people he loved. They needed to know he was OK and happy. He wanted them to be happy for him, but that might be a lot to ask. The chief conductor finally announced their arrival at Oxford Station.

A young airman, about Brian's age, saluted him as he walked down the platform. "May you be Pilot Candidate Drummond, sir?" he asked.

"Yes."

"Very good, then. I am Aircraftman Jamison, and I'm here to fetch you for the flying school."

The salute from Aircraftman Jamison, the lorry driver, was a bit odd for Brian, but after all, he was in the uniform of a candidate officer of the Royal Air Force, even if he was only a fledgling pilot without his wings, yet.

The young driver remained quiet until Brian began to ask questions about different features of Oxford. Jamison picked up the accent, acknowledged their common age, and opened up to Brian offering a running commentary as if he was a practiced tour guide.

The streets of Oxford, with its famous university of higher learning, did not appear to be substantially different from London although the stones of the buildings were darker, heavier to the eye. The streets and walks were narrower with even less traffic. The pace of the city was decidedly slower than the capital. Oxford was also smaller, much smaller, than London. It took a mere few minutes to meander through the narrow streets, the abundance of large, green, leafy trees and the stimulating smells of local bakeries to reach the green fields and hedgerows. The gentle rolling hills and the narrow, snaking roadway made the distance seem much longer. The rich, redolent fragrance of plowed fields, deep brown earth and things growing brought a broad smile to Brian's face.

"This is great," Brian said.

"What is, if I may ask, sir?"

"The smells. The smells of the fields, farming, it reminds me of my home."

"Where might that be?"

"Kansas."

"Oh, yes, indeed, tornadoes, Toto, the Yellow Brick Road, and all."

"That's right."

There was no sign at the gate across the opening in the high, thick hedge. Two armed guards stood holding their weapons at the ready and did not move until they recognized the driver.

Stopping at the gate, Aircraftman Jamison announced, "Another almost, here in search of his wings."

After looking into the car to confirm the rank of the passenger, the corporal of the guard requested, "May I see your orders, sir?"

Retrieving the multiple sheets of paper from the inside pocket of his uniform jacket, Brian handed them to the guard. The quick reading and confirmation of the proper stamps was all the corporal needed to see. "Thank you, Mister Drummond. Welcome to RAF Brize Norton."

As they drove past the gate, Jamison explained, "No disrespect meant, Mister Drummond."

"What do you mean?"

"Calling you an 'almost.' It's sort of a tradition here at the training wing."

Brian laughed. "No problem."

"Very good, sir."

The base was clearly an air force installation. A wide variety of men and even a few women dressed in the distinctive blue uniform of the RAF, or the dark blue coveralls of an aircraft mechanic moved from building to building with a sense of purpose to their walk. Most of the buildings were made of wood board, painted medium green with simple shingled roofs that gave them a rustic, rural appearance to Brian.

Aircraftman Jamison assisted Brian with the check-in procedures, showed him the essentials like the Officer's Mess, and then deposited him at the Student Pilot's Residence. "With your permission, sir, I'll retrieve you at half past eight for the remainder of your check-in," announced Jamison.

Monday, 19.June.1939
RAF Brize Norton
Brize Norton, Oxfordshire, England

"Good morning, sir. I'm afraid I'll need to take you directly to Wing Commander Henry's office, sir."

"OK."

"A word of advice, if I may, sir. The wing commander is not a very friendly person, but he is good. So, I might suggest, choosing your words carefully and keep your answers short."

"Thanks."

The single level, wooden buildings gave way to three large, concrete and steel, hangars beyond which lay the flight line. The actual airfield looked essentially the same as every airfield Brian had seen. Airplanes parked in rows. A control tower with its radio antennae sprouting from the flap roof over the prism like, flat faced windows of the octagonal observation room. They were all the same.

Aircraftman Jamison stopped near the door leading into the building which was the base of the control tower. The base commander's office was probably in this building, Brian told himself, as Jamison motioned for Brian to follow him into the building.

The large room inside the door had an enormous standing desk like a bar along with the maps and chalk boards on the walls. It had to be the operations room. Several pilots as well as support personnel carried on the business of the flight training operations. Everyone looked toward Brian as he entered making the young man somewhat uneasy that most of these new eyes were watching him.

"Fresh meat for the grinder, do you say there, Jamie?" one of the pilots shouted.

The speaker was a medium size man with brownish red, wavy hair and a bright red, full handle bar mustache. He was wearing the uniform of an RAF flight lieutenant. Probably one of the instructor pilots, Brian told himself.

"Yes, sir, Mister Morrison," the Aircraftman responded. "This is,

Mister Brian Drummond, everyone."

"Welcome." "Good to have you with us, Drummond." "Welcome aboard." The greetings flooded the room with a common friendliness.

Brian raised his hand and said, "How d'ya do."

"Ah ha," exclaimed Flight Lieutenant Morrison. "We have a Yank amongst us and a young one at that, I suspect."

"The RAF can't be that desperate we're importing throttle jockeys from the colonies," some anonymous taunter said.

"Now, now," responded Morrison, "enough of this imperialist tripe."

"This way," Aircraftman Jamison quickly interjected motioning down the hallway.

The hall was not wide nor long with a couple of windows on the left, two doors along the right wall and one door at the end with the words, WING COMMANDER JASON G. S. HENRY, OBE, painted on it.

Jamison stopped to look at Brian, his eyes asking if Brian was ready for an ordeal. Brian sensed the unspoken question, checked his uniform tunic and tie, and then nodded his head. The enlisted man knocked twice rather loudly.

"Enter," came the booming reply from the other side of the door.

Jamison opened the door, took a step into the office, stomped his boot and announced, "Pilot Candidate Drummond, here to see you, sir."

"Very well," responded Wing Commander Henry. "That will be all, aircraftman," he added as if those were code words for the enlisted man to let the new man into the office, then leave shutting the door behind him. The commander was a slight man, completely bald with a full mustache and an appearance of age about him. The expression on his face as Brian entered the room was almost a scowl as if he was very angry and trying to contain his emotions.

As Brian walked smartly toward the senior officer's desk, he kept his eyes on his superior, stopped one yard short of the desk, stomped his boot and saluted as he had been taught by Group Captain Spencer. A rather nonchalant salute was returned. Wing Commander Henry held out his left hand, palm up, without saying another word. Brian had to guess what he wanted. The orientation of his hand suggested he wanted something placed in it rather than to shake Brian's hand. It took a few seconds too long for Brian to figure out he wanted his orders.

"You're a smidgen slow there, young man," he said with a very stern, cold voice. He read Brian's orders quickly, evaluated the stamps for authenticity and then placed the orders to the side of the desk. "So, you are a bloody American and you want to be a pilot, ay?"

"I am a pilot, sir," Brian said instantly realizing that was probably not the best thing to say.

"I'm afraid I shall have to burst your bubble," he said with some increased intensity to his voice. "You are not a pilot until I say you are a pilot. Is that clear?"

"Yes, sir. I just meant I am a pilot and I want to be an RAF fighter pilot."

"Well, you can want all you care to, but until I say you are a pilot, you are just another sodding bloke."

"Yes, sir."

"If I read your rather short file, here, you are just eighteen years old."

"Yes, sir."

"And, you are a sodding American."

"Yes, sir."

"What the bloody hell are you doing here?"

Brian was not quite sure how to answer the question, if it really was a question. The wing commander just stared at the young candidate obviously expecting an answer. Brian could feel the beads of sweat rolling down his backbone.

"I want to be a fighter pilot in the RAF, sir."

"That may be, but how does a child Yank show up in my office as a bloody officer?"

Again, Brian was not sure how to answer the question. "I don't know, sir."

"You don't know, ay. Isn't that charming. It must be divine intervention, I suppose." He paused staring at Brian as if he expected some reaction. He got none. "You must have a benefactor. Someone important, to receive a commission without completing university. You didn't complete a university, did you?"

"No, sir."

"Then, who's your mentor?"

"Excuse me?" Brian said not knowing what the word meant.

"Who got you here?"

"Group Captain Spencer, sir."

Wing Commander Henry leaned forward in this chair as if he was getting ready to lunge across the desk. His eyes narrowed. "You don't say."

"Yes, sir."

"Well, well, the nephew of our infamous MP."

Brian had heard the term, MP, before although it took him a few more seconds to remember the initials meant, Member of Parliament. The wing commander had not asked a question and Aircraftman Jamison's words flashed back to him.

"And, how did you come to know, Mister Spencer?"

"Through my friend."

"Who is?"

"Malcolm Bainbridge."

"I know that name," Henry said pausing to consider his memory. "Do you mean, the American, Malcolm Bainbridge, who flew with the Royal Flying Corps in France during the Great War, The Malcolm Bainbridge, holder of two DFC's?"

"Yes, sir," Brian answered somewhat surprised. He knew quite well that DFC stood for Distinguished Flying Cross, one of the highest awards for heroism in the air.

"I'll be damned, boy. You are well connected." Henry stopped to simply stare at Brian making the younger man very uncomfortable. "Well, now," he continued leaning back into his chair talking through clenched teeth, "let me tell you something. This is my training wing and no one, not even sniveling little boys of big MP's, gets through this course without measuring up to my standards."

"Yes, sir."

"Do you understand?"

"Yes, sir."

"Well, then, tomorrow will be your day of reckoning. You and I shall fly your first flight together."

"Yes, sir."

For the first time since his initial flight with Harry Johnson, the barnstormer pilot when he was 9-years-old, Brian was not looking forward to flying. He knew from the adversarial tone of the wing commander, his first flight in the RAF was going to be a trial. Brian also knew he had to get past Wing Commander Henry, or his dream would vanish in a flash.

"Aircraftman Jamison will get your flying kit and give you an instruction checklist that is intended to prepare you for this training. I will see you at my aeroplane at oh nine hundred hours tomorrow and we'll see whether you have the guts to be a fighter pilot."

"Yes, sir."

Wing Commander Henry pushed a button beside his desk telephone. Aircraftman Jamison entered the room a short, few moments later. "Take Mister Drummond to get his flight kit and complete his preparations."

"Yes, sir," answered Jamison.

"You're dismissed, Drummond. I'll see you again shortly," Wing Commander Henry said. A thin smile, almost a sneer, came to his face. "I can't wait to see just how good your friends think you are."

"Yes, sir."

A profound sense of relief descended upon him as he left the commander's office and the door closed behind him. Brian felt the strongest urge just to get out of the building as if the extrication would resolve the feelings of confusion, regret and a little anger.

"Welcome to the grinder," Flight Lieutenant Morrison said in a near laugh as the young pilot left the operations building.

Silence was all Brian wanted as Jamison drove him to the supply building where he was issued his flight kit, the equipment he would use during his flight training. Aircraftman Jamison made several attempts to strike up a conversation with no response from Brian. The words were lost on Brian as Jamison completed a tour of the base showing Brian the officer's mess where he was supposed to eat, the other administrative buildings, the Wing Commander's airplane that made his stomach turn over, and lastly the student pilot's quarters.

The remainder of the day was a blur to Brian as he struggled with the emotions around his meeting with Wing Commander Henry. He packed his kit bag and suitcase, ate a small meal, met several of the other student pilots without seeing their faces, hearing their words, or remembering their names. Before he was safely in the arms of sleep, everyone seemed to have heard about Brian's meeting with Henry. Condolences were the majority wishes offered.

Chapter 5

> To test man, the proofs shift.
> -- Robert Browning

Tuesday, 20.June.1939
RAF Brize Norton
Brize Norton, Oxfordshire, England

FORTUNATELY for Brian Drummond, a good, restful sleep brought a clear, focused mind. Several of the other students reiterated their cautions and advice from the previous night. Brian had not heard them the first time.

Each one of his colleagues told their own story about their first flight at Brize Norton. Most had flown with Wing Commander Henry. Every student who indicated a desire or an interest in flying operational fighters without exception got special treatment from Henry. The common thread seemed to be a purposeful attempt to filter out any candidate who could not take the rigors of aerial combat.

Brian listened. No one asked him if he had flight experience or if he had, more specifically, aerial combat maneuvering experience. As Malcolm had recommended, he decided to let his actions speak for him. Brian also absorbed the more practical information like flight attire, procedures, protocol and rules. Helpful hints about how to survive, and what to expect did not pass by Pilot Candidate Brian Drummond.

The picture was quite clear. This was not going to be an easy flight. Brian decided early he would not eat breakfast just in case. A few crackers and some tea would hold him over until after the trial. He checked his uniform one last time in the mirror, grabbed his leather flight hat and gloves along with his goggles, and walked to the flight line.

Wing Commander Henry's bright yellow, diminutive de Havilland DH68 Tiger Moth biplane sat alone in front of the operations control tower. Arriving 15 minutes early enabled Brian to talk with the leading aircraftman for the commanding officer's airplane, Sergeant Jones. A different version of the first flight saga surrounding Jason Henry complemented the earlier renditions. The objective was clear. The mythology of the first flight helped Brian focus on the skills, techniques and concentration needed for the impending event.

"Good morning to you, Sergeant Jones," announced Henry, as he walked briskly toward the aircraft.

"Morning, sir."

"I see the other candidates have helped you with your kit, Mister Drummond."

"Yes, sir."

"Are you ready for the reckoning?"

"I guess so, sir."

Henry stopped, faced Brian directly with his leather flight helmet and both fists perched on his hips. "You do not guess in this business, Mister Drummond. Either you know or you do not. Guessing simply gets you killed. Is that clear?"

"Yes, sir."

"Then, let us try again. Are you ready?"

"Yes, sir."

"We shall see. You take the front cockpit. If you feel the urge to retch, you will do so into your uniform, not over the side or in the cockpit. Sergeant Jones does not take kindly to cleaning up after an 'almost.' If I get wet from your weak stomach, you will not fly again. Is that clear?"

"Yes, sir."

"Then, let us get to it."

Sergeant Jones helped Brian with his parachute and strapping into the front seat although he really did not need the assistance. The extra attention did ease the tension slightly. As Brian stepped through the normal flight preparations, his thoughts captured the contrast between Malcolm Bainbridge and Jason Henry. Their styles could not be more different. Brian reminded himself, this flight was just another obstacle to be surpassed, but he wished it was Malcolm doing the evaluation. To be so close and not make the objective brought its own strain to this event.

Wing Commander Henry performed all the procedures himself without any participation by Pilot Candidate Brian Drummond. As Malcolm had taught him, Brian followed each control input with his hands and feet lightly on the controls. Once at altitude, the throttle was retarded to idle as Henry pulled back gently on the stick slowing the aircraft. Stalls, Brian told himself with a smile no one could see.

The shakes, shudders and gyrations of an airplane at stall were quite familiar to Brian. The Tiger Moth stalled like most of the biplanes he had flown, nothing spectacular.

With no detectable reaction by his student, Jason Henry took the inevitable next step, another stall. Only this time as the characteristics of the stall presented themselves, Henry added full rudder. Brian instantly recognized the control inputs to induce a stall spin. The opposite wing dropped sharply followed quickly by the nose. Henry let the aircraft settle into a fully developed spin. The world rotated rapidly with the nose well below the horizon. After about ten turns, Brian felt the proper anti-spin controls go in. The aircraft returned to stable flight without any complications.

With his head spinning and the recognizable waves of nausea coming to him, Brian felt Wing Commander Henry enter another spin in the opposite direction this time adding opposite aileron to aggravate the spin entry and make the aircraft gyrate more violently. Again, the recovery was normal.

His body reacted strongly to the rapid motions and the fact that he had not flown in more than five weeks. He sucked in several very deep breaths trying to clear the lightness in his stomach and the 'sweats.' He knew the nausea would pass and not come back as his body adapted to the motions of flight. Brian had done these maneuvers many times before. The process of adaptation would soon relegate the body responses to the bag of experience.

After several types of spins and still no detectable reaction, Jason Henry flew a continuous series of rolls, loops, barrel rolls, Immelmann turns and other aerobatic maneuvers. Brian's nausea persisted although there were no signs of worsening symptoms. The confidence and anticipation returned comfortably to Brian as the flight progressed.

At the conclusion of what was probably 40 minutes of continuous maneuvering, the wings leveled with the airspeed stable at 80 miles per hour. "Take the controls, Mister Drummond," shouted Henry.

Brian did as he was told keeping the aircraft easily straight and level. The aircraft required very slight, continuous inputs to hold the condition. Brian wondered what was going to happen next? Was everything OK in the back seat? He waited. As he waited, the last vestiges of his nausea disappeared.

"You can maneuver however you wish," came the command from the rear cockpit.

"Yes, sir."

The young aviator took the cue to repeat some of the maneuvers demonstrated by Wing Commander Jason Henry as well as a good many additional maneuvers he used to perform with Malcolm Bainbridge like high scissors, tuck-under-rolls and spiral descents.

"Take us back into land."

"Yes, sir."

Fortunately for the pilot candidate, the air was crystal clear with only an occasional puff of a small cumulus cloud. Their location relative to the airfield was easily determined. The most difficult task was establishing the wind direction and therefore the landing direction into the wind. The wind was light and there were no trails of smoke or dust. Eventually, a small gust of wind passed over a pond rippling the water and providing a general wind direction. Traffic in the vicinity of RAF Brize Norton was relatively well defined with several aircraft in an oval landing circuit. Brian remembered what the other pilots told him about the entry into the landing pattern and followed those procedures. He figured Henry would let him know if he was doing something wrong.

The landing was fairly easy with the typical few bounces as the aircraft hit the uneven grass. He taxied up to the same spot at the base of the control tower and turned the aircraft around as he had done so many times before. With no further direction, Brian waited until he became concerned about whether he should switch off the engine.

Turning to look back at Jason Henry, he asked, "Should I shut down?"

"Oh, by all means, please do."

The tone of Henry's response did not bode well for what was ahead. What had he done wrong? Had he done too much? Did Wing Commander Henry think he was showing off?

Henry was out of the aircraft and walking into the operations building before Brian finished unstrapping. Brian looked at Sergeant Jones for a clue and received a simple shrug. Following Henry into the operations building, he found his evaluator talking to the man introduced as Flight Lieutenant Morrison.

"In my office, Mister Drummond," said Henry motioning toward the hallway and leading him. Henry waited for Brian to enter, then shut the door behind him. Without further movement, he began, "Why didn't you tell me you had flown before?"

"I did, sir."

"Well, well, we must have a player here," he said moving behind his desk and sitting down. He motioned for Brian to take a seat in front of him. "I presume you learned to fly with Malcolm Bainbridge."

"Yes, sir."

"How much time do you have?"

"About eight hundred hours, sir."

The discussion turned to a new level as one pilot to another comparing notes and experiences. Jason Henry had flown in the Great War, also. They reviewed all of Brian's experience from his training, to the courier service and the airmeets along with his navigation and weather flying experience.

"You clearly do not belong here. I'm going to accelerate your primary training to make sure you are a proper officer in His Majesty's service. You will fly with Mister Morrison in order to cover the way we do things here and the expectations of you within the RAF. What do you want to fly?"

"Spitfires."

"Of course. I will be most happy to endorse your dossier after you have completed the abbreviated course. This country will shortly need the skills of pilots of your caliber."

"Thank you, sir."

"Don't thank me, Mister Drummond. This is going to be a bloody awful row with the Nazis."

"Yes, sir."

"On your way out, please ask Mister Morrison to come in here. You are dismissed."

"Yes, sir," responded Brian before turning to leave. Mister Morrison looked up from the operations desk.

"I assume the old man is ready for me."

"Yes, sir."

"You may have a seat over there to wait for me."

Brian sat down in the designated chair. The flight lieutenant walked casually into Wing Commander Henry's office leaving the door open slightly. "What did you think?" asked Morrison.

"Jeremy, this kid is one of the best pilot's I've ever flown with."

"Except me, right?"

"No. He's better than both of us. Our job is to polish this diamond. I want to accelerate him."

"How much?"

"Two weeks."

"No problem, if he's that good."

"No, Jeremy. I mean I want him to the OTU [Operational Training Unit] in two weeks."

"Jesus wept, are you kidding?"

"Certainly not. This lad is something. He has about eight hundred hours mostly with Malcolm Bainbridge whom you may not know, an American who flew in the RFC . . . two DFC's as I recall. He's a natural, and we are not far away from another world war. We need to ensure young pilots like him have all the tools to survive and win."

"Two weeks, you say?"

"Yes."

"Then, I should get busy."

"Take him out as soon as you can. Arrange for Huntington to takeoff after you and meet you for some mock combat. Let us see what this young man can do. I suspect he has learned well from his tutor."

"This Malcolm Bainbridge?"

"Yes."

"Then, we will give him a good taste."

Jason Henry turned his attention to the mounting paperwork on his desk. Flight Lieutenant Jeremy Morrison returned to the operations room.

"Mister Drummond, are you able to fly another flight?"

"Yes, sir," responded Brian wondering what was going to happen next.

Brian was instructed to proceed to another Tiger Moth, jump into the front seat and prepare for takeoff. Morrison would follow shortly. With Brian enroute to the assigned aircraft, Morrison issued the appropriate instructions to Pilot Officer David Huntington, a highly regarded fighter pilot doing his stint in the Training Command.

The setup followed the exact script. Morrison handed Brian a map under the pretense of a navigation exercise. Huntington took off ten minutes later and rendezvoused with Morrison and Drummond maneuvering to a firing position behind them.

"Enemy on your tail," shouted Morrison.

Brian instantly pulled back hard on the stick, rolling and then diving straining his neck and head to check his tail and confirm the adversary. Despite Huntington's anticipation of a maneuver, the rapid combination surprised him as well as Morrison. In a relatively short span of time, Brian was the hunter and his attacker was gyrating to shake him.

Ten engagements were flown between the two aircraft. The experience of the veteran RAF pilot and the lighter weight with only one pilot gave Huntington the advantage. Morrison judged five wins to Huntington, three to Drummond and two draws. The fact that Brian Drummond won any of the engagements was testament enough to his skills.

The return to base was uneventful. The debriefing concentrated upon the flight experience of both pilots. Brian had done well, but he sensed a relationship of mutual respect had begun that afternoon.

—

Monday, 26.June.1939
RAF Brize Norton
Brize Norton, Oxfordshire, England

THE burden of the expectations of others made the task of assimilation far more difficult than would otherwise be the case. Each day was filled with lessons on military etiquette, customs and procedures in addition to the practical aspects of air traffic control and aerodynamics. Brian received the extra attention of a thoroughbred. He was certainly not the only pilot candidate to be so recognized, but he was the only American so far. His nationality attracted both positive and negative elements.

The flying lessons focused on the practical application of skills to the operational characteristics common to the Royal Air Force. Just being a great stick and throttle man could not guarantee a great operational fighter pilot. Each pilot had to function within a specific system that required all the pieces to work in a coordinated, purposeful and focused manner. Some of the procedures were not intuitively obvious to Brian as well as other fledgling pilots.

As a relatively naive young man from the Central United States, he was also an easy target for those less capable individuals who wanted to show their superiority through practical jokes intended to demonstrate his naiveté. While his flying skills were recognized, his stature as an RAF officer fell a bit short. The refinement, polish and social skills accepted as standard for British young adults were either missing or immature in Brian.

Several students and one instructor took to Brian maybe due to his nationality, or strange accent, or his even temper and quiet nature. Most notable among Brian's growing circle of friends at RAF Brize Norton was Flight Lieutenant Jeremy Morrison. Although his stature was quite near the average for RAF pilots, medium height and build, his appearance was certainly contrary to the norm. His full head of wavy, reddish brown hair along with his square face, sharply defined bones and the rosy pink cheeks made him standout like Michelangelo's 18-foot statue of David. His sky blue eyes and full, prominent lips seemed to make him popular with the ladies. His skills and comradeship made him well liked within his profession. The consummate fighter pilot, Jeremy Morrison not so gracefully took his assignment to the Training Command and looked forward to his return to Fighter Command. The tasking from Wing Commander Henry to tutor and shepherd the young American wannabe gave him a clear objective. The fact that it was easy to like Brian Drummond made the task even more enjoyable. Morrison pushed Brian very hard, stretched his limits of fatigue, concentration and skill.

This particular Monday blurred before the non-stop series of flights, ground lessons and military instruction. They worked in relays through the weekend with no relief in the tempo for Brian.

"You must concentrate on the marker points, the cues," admonished Morrison.

"I know," responded Brian.

"You may know, but you are simply not performing. You could be the best solo pilot in the world, but if you can't master formation flying, you cannot function in Fighter Command, and what's more, you will be a risk, a burden, to your flight and squadron . . . simply put, people get killed."

"Yes, sir."

"What is the problem, then?" Morrison asked while removing the remainder of his flight kit and leaning against the lower wing of the Tiger Moth.

"I feel like I'm going to run into the other airplane."

Jeremy Morrison laughed, walked in a small circle, and then faced Brian directly. "We all feel like that at first. Some get over it, some never do. You do everything near perfect except close formation flight." Brian stared at Morrison not knowing whether he should respond or not. Flight Lieutenant Morrison turned pensive considering how to approach the solution. "Do you remember when we flew the mock combat on your first day?"

"Yes, sir."

"Once you got on Huntington's tail, you flew close, and you wouldn't let him shake you."

"That's different."

"No, it isn't! It is exactly the same except you are flying off his wing. Brian, the secret to formation flight is, detecting the relative motion. The closer you are, the easier it is to detect any relative motion. If you move in, closer to the leader, you will find the station keeping task much easier, as well as any maneuvering."

"But, if I do that, I can't look anywhere else. I can't keep up with navigation. I can't look for enemy fighters. I can't do anything except fly position."

Morrison considered Brian's observation. "You are correct, but if that is what your leader wants, then that is where you must be."

"It doesn't make sense," Brian protested.

"That may be, but that is the way we do this business."

"I've never had to fly so close to the wing of another aircraft before."

"Your instructor did a very good job teaching you to fly, but you must learn to operate with a team and follow the directions of your leader, or you will never make a good fighter pilot. Brian, you are one of the best natural pilots I've ever flown with, but I am dead serious . . . if you don't master close formation flight, whether it is right or wrong, you will not make it to Fighter Command."

Brian thought hard about Morrison's words. The sincerity amplified the meaning. He knew he had no choice. "Can we go back out and try it again?"

"You've already been out twice. I think that is enough for one day."

"Please, sir," begged Brian.

The instructor pilot weighed the reinforcement against the fatigue.

"As you wish. Let me see if we can find a leader."

Brian walked around the diminutive biplane trainer reviewing each of the marker points established by visually looking across the wing to positions on the fuselage. Each marker point established the proper relative angle between two aircraft in proximity to one another. The only other piece of judgment was the distance. Maybe Morrison was right, he told himself . . . the closer you got to the leader, the easier the station keeping became.

"Right, then," Morrison said upon his return. "We'll have a leader. Huntington is going to do a test flight on one of the aircraft in about twenty minutes. Let's go fly around until he gets up, then we will rendezvous with him and run through all the formation maneuvers."

"Yes, sir."

RAF Brize Norton sat in a soft bowl in the gently undulating terrain about 15 miles due west of Oxford. The farming country was not appreciably different from his native Kansas with one significant difference – most lines and features in Kansas except rivers and streams were straight, orthogonal arrangements. Nothing seemed to be straight in England. The unique character of the terrain, small villages, roadways, hedgerows and quilt of fields made navigation significantly easier than in the American Midwest. It was truly a beautiful country.

"Here comes our leader," announced Morrison from the back seat.

Brian did not need to ask for directions. His quick scan located the small aircraft heading toward them. He knew what was expected as he turned to position his machine for a proper rendezvous turn and join up. Maneuvering around another aircraft had been a quickly acquired skill from his flying with Malcolm Bainbridge. It was the close in, called parade formation, flying that bothered him the most. As the distance between the two aircraft rapidly decreased, Brian reminded himself of Jeremy Morrison's words.

Get close. Concentrate. Use the marker points to quickly pick up any relative motion and make the appropriate corrections to maintain the proper position. It all sounded so easy in talk or thought. Somehow he had to get over his fear of colliding with the other aircraft. He simply had to overcome his fear.

As he settled into position behind Huntington's Tiger Moth, he immediately recognized the tightening in his arms and legs. He struggled to maintain position. The throttle and stick movement grew bigger. Move in, he told himself, move in. Gradually, he forced himself to move his aircraft toward the lead aircraft. As their wings began to overlap, he was certain they were going to collide, and he immediately backed off.

"I've got the controls," shouted Flight Lieutenant Jeremy Morrison.

The instructor pilot moved the aircraft closer to the lead. Brian felt himself pushing back into his seat. He followed Morrison's control inputs. As he said they would be, the movements were smaller and less frequent, smoother. Morrison settled the aircraft into position slightly behind, stepped up, and with half a wing overlap. The aft vertical, wing support, wire brace fastener intersected the forward exhaust port. Picture book position cue.

Huntington performed several climbing and diving turns allowing Morrison to demonstrate the station keeping skills for Brian. It did seem so much easier with Flight Lieutenant Morrison flying the aircraft.

"This is where you should be," shouted Morrison. Brian nodded his understanding. "You take control and hold us in this position."

The aircraft jerked and shuddered as Brian took control instinctively wanting to back off.

"Stay in position!"

Brian told himself to concentrate just as he had for every difficult task in his flying experience. Relax, he could hear Malcolm telling him, just relax. Brian tried to do what he knew had to be correct. Ever so slightly, the aircraft began to settle down. The machine seemed to naturally want to stay in position. Brian's muscles responded to the commands to relax. Hey, Brian said to himself, this really works.

The maneuvers began to increase in complexity. Brian stayed in position. The urge to fall back slowly began to subside as Brian's confidence in the technique grew.

"Let's break it off and head for home," shouted Morrison.

The return to RAF Brize Norton brought a smile to himself. For the first time since being introduced to close formation flight, Brian began to feel better, almost good.

"Not bad," Morrison said as they stood on the ground. "I think you may have crossed the barrier. We'll give you a good run through tomorrow, but I think you've got it."

"It really works, sir. Being in closer really works." Brian's excitement brought a smile from Jeremy Morrison.

"Yes, it does. Just remember what I told you."

"Yes, sir." Brian acknowledged his instructor, then remembered some of his misgivings. "It still doesn't seem right for combat, though. You have two aircraft and two pilots, but only one set of eyes scanning for the enemy. What if you get into clouds and have to break up? What does the wingman do about navigation?"

A sharp, quick laugh greeted the common observation. "Quite right, I'm afraid. However, the profession of aerial combat to which you aspire is characterized by moments of frenzied, frantic mortal action followed by periods of serious apprehension wondering if you'll make it back home. You must master the art of rapid reorientation with your surroundings and making sure you are where you need to be. You have only two methods of navigation. One, the RDF controllers will get you to the spot in the sky they want you to be and help you get pointed in the right direction back. Two, your brain. You will most likely have no one else to assist you. You will be dependent upon your map, the clock, the airspeed indicator and compass, as well as any appreciation of the terrain below you if you can see it. You would be surprised how quickly you get reoriented when your life depends on it."

"I suppose so," Brian said.

"Righty ho, then I think we both could use a beer. Shall we retire to the mess?"

"Sounds great."

The two pilots compared observations from the day's breakthrough flying. They both felt better than they had in the morning. Brian sensed he had overcome the final obstacle while Jeremy knew he would be able to fulfill the objective given him by Wing Commander Henry. For the first time since their introduction, the two pilots told aerial stories from different locales, but the unique common experience of aviators.

Monday, 3.July.1939
Headquarters, Secret Intelligence Service
No.21 Queen Anne's Gate
Westminster, London, England

ADMIRAL Sir Hugh Francis Paget Sinclair, KCB, the aging and ailing Director-General of the Secret Intelligence Service and only the second person to hold the pinnacle position of 'C' inside SIS, occupied a large office not particularly unlike most men of comparable position and responsibility within His Majesty's Government. Dark, elegant mahogany paneling accentuated the well-stocked book shelves. The medium green, official carpeting brought some lightness to the otherwise dark office.

The conference table was prepared as it always was by his long time, devoted assistant. This particular Monday looked different from many previous Monday's. The simple message – "matter of ultimate importance" – from Commander Alastair Denniston, his chief of the top-secret Government Code and Cipher School at Bletchley Park otherwise known as Station X, peaked his curiosity. The fact that Denniston sent the message from headquarters of French intelligence service made the anticipation more pronounced.

The two short buzzes followed by the knock on his closed office door announced the arrival of one of his two guests. First to arrive was his counterpart at the Admiralty and equally aging Director of Naval Intelligence, Vice Admiral Sir Geoffrey 'Jumper' Pike.

"Good afternoon, 'C,'" said Pike.

"Good afternoon to you, Geoff. Thank you for coming over."

The two senior intelligence officers shook hands then sat in two of the four chairs around the small table with a newly refreshed tea service. Sinclair poured tea as they discussed family, handed a cup to Pike, and then added some sugar and cream to his cup. As he stirred, Pike did the same.

"Denniston is expected shortly."

"Excellent. I hope he brings some good news."

"Indeed."

"What on earth was doing in Paris?"

"To be honest, we are not sure. The French asked for him specifically on a matter of utmost urgency. I am just as curious as you, I'm afraid."

Pike nodded his head. "How do things look to you?"

"We need some other opinions, but from my perspective, not good, I must say. The Nazis have been massing their troops, a full range of *Wehrmacht* units including several *SS Panzer* divisions that we know of, along the entire Polish frontier principally in Germany but several units are positioning in Czechoslovakia as well, and the hooligan activities in the Danzig enclave are becoming more visible and violent."

"What type?" asked Pike.

"Simple vandalism of merchant shops to rapes and beatings. The level of violence has increased markedly. The victims are no longer just ethnic Germans. Several notable anti-Nazi Poles in the enclave have been victimized as well."

"Tragic."

"Surely," added Sinclair. "Poland is next and not too far off, I should think."

Both men stared at their tea cups or some other far more distant place.

"Both *Deutschland* and *Graf Spee* have put to sea in the last few days," Pike offered without being asked. "By themselves, not a good sign. The real marker for us will be the U-boats. We have indications many are preparing for sea, loading provisions, ammunition, torpedoes and such, but so far none have set sail. If hostilities are to commence, a goodly number of the U-boats will make for the open ocean roughly a week prior. So, we are keeping a close eye on them."

"Does the PM know?"

"Yes. The First Lord briefed him day before yesterday."

Another doublet of the buzzer and door knocks. Commander Denniston joined them. Greetings, chit-chat about the warm weather, and replenishment of tea were dispensed with promptly.

"What have you for us, Alastair?" asked Sir Hugh getting right to the point.

"Well, to be candid, I must say it was a bit of a shocker. As you recall, Sir Hugh, this past January we met with our counterparts in Polish intelligence. We were aware they had some breakthrough work on the German codes, but we did not know to what extent." Denniston paused to look each of the senior men in the eyes for some reaction or questions. Stone cold. "Well, let me just summarize the content of this meeting by saying the Good Lord above, may he have mercy on our souls, has chosen to bless us. But, more to the point, thank God for the English Channel."

"Yes, yes, Alastair. Enough of the melodrama. Get on with it," said Sinclair.

"The Poles anticipate war with Germany just as we do. They made arrangements with the French to move their cryptanalysis group to Paris, *in toto*. Furthermore, they fully disclosed to both the French and us the extent of their crypto work. There is so much to say, and I want to say it all at once."

"Take your time," Sir Hugh added.

"Yes, well, here we go. The Polish mathematicians in their crypto-section have performed some absolutely magical work in breaking down the key sheets for the German Enigma device nearly a decade ago, as we learned. They managed to piece together a functional knowledge of the device. The critical element for them actually came from a French windfall. A German embassy civil servant apparently had a crisis of conscience as Hitler and his Nazi thugs were coming to power. This agent provided the French with the service and operations manuals for Foreign Office Enigma device."

"Dear God above," Pike interjected.

"Yes, yes, but the best is still ahead. The French, unbeknownst to us, knew the Poles were having some success, although they claim they did not know to what extent. Nonetheless, they provided copies of the manuals to the Poles. With those manuals to fill in the last remaining gaps in their knowledge, the Poles constructed a working copy of Enigma. More specifically, they produced six copies. And, they will keep two for their operations in Paris;

they gave two to the French, and two to us."

"Dear God above," repeated Pike.

"We agreed upon a curriculum to expand the Polish work. It seems the Germans have improved the device by adding another rotor for a total of four. This added complexity is beyond the capacity and resources of the Poles. We also have a copy of the mechanical computing machines, they call Bombes, which they created to breakdown the key list. Probably of even greater significance, this last spring they acquired an actual device from an unsuspecting SS unit that confirms precisely the reconstructions they derived."

"You mentioned a curriculum," Sinclair interjected.

"Yes. Our task, first and foremost, must be to protect this capability at all possible costs."

"Agreed," the two senior officers said in unison.

"Second, the machines are of little practical value without the key lists for any given day or message. We have not refined our calculations, but the four-rotor device has something close to a 100 million million combinations. Now, while that statistics appears rather daunting, the Poles have managed to identify some very key tricks to establish a given key list."

"Which may very well change once or more a day," observed Pike.

"We could be chasing our tail," Sinclair added.

"Indeed. However, from what I saw, the more we use this system, learn little bits of how they use it, and find some successes beyond what the Poles have already done, we shall be far more adept at the codebreaking process."

'C' nodded his eyes as his eyes flashed. "As you say, we must first secure this capability. How many know of this at the moment?"

"On our side . . . you, Admiral Pike and me. Our device and the materials I carried are safely locked in my personal safe at Bletchley. No one else has seen them. On the French side, from what I gather, a comparable number. For the Poles, perhaps half a dozen or so."

"We shall keep the number very small for the time being," said Sinclair. Looking to Pike, "I should think Bletchley is better place for this work than your Room 40 at Admiralty, wouldn't you say, Geoff?"

"Indeed. I dare say none of us needs to see how the sausage is made . . . just eat the exceptional product."

"Well said," Sinclair responded. "So, we will need a security specialist."

"I would suggest Wing Commander Fred Winterbottom at the Air Intelligence Branch. He has exceptionally good relations with MI-5 and Scotland Yard, and we need an air force contact point."

"I know Winterbottom. Good man. Geoff?"

"I as well . . . excellent choice."

"Then, we are agreed. Alastair, if you would be so kind, bring Fred into our little circle, have him up to Bletchley for an independent assessment, and the two of you work out to ensure the fullest possible security measures are put in place immediately. If you have any problems whatsoever, you are to contact me immediately. Understood?"

"Yes."

"Now, the last order of business might be classification. Any information associated with this device, no matter how remote, and specifically any material derived from the Enigma system will be coded – MOST SECRET-ULTRA. No one will be given access without the expressed written approval of both myself and Admiral Pike. Wing Commander Winterbottom will be the keeper of the master list."

"What about the PM?" asked Denniston.

"Or, Winston?" added Pike.

"For the moment, I do not think any politician should be trusted with this secret. The risks are simply far too great at this critical juncture."

Sir Geoffrey pondered the statement for several seconds, decided not to remind "C" of Churchill's assistance in the acquisition of the unit, and then nodded his head in agreement. Denniston nodded his head as well.

"Then, we are agreed?" asked Sir Hugh. Again, nods. 'C' stood and extended his hand to Denniston. "Congratulations, Alastair. Well done. Well done, indeed."

"Not my doing, I'm afraid, 'C.' The blessings of the Good Lord have come to us."

"Nonetheless, well done."

"Yes," added Pike, "truly manna from heaven."

Chapter 6

*Love is a canvas furnished by Nature
and embroidered by imagination.*

                                                                -- Voltaire

Friday, 7.July.1939
RAF Brize Norton
Brize Norton, Oxfordshire, England

"WELL, young man," said Wing Commander Henry, "you've managed to disprove all those who were convinced you would not be able to cope with the military environment. In addition, you've performed the prescribed flying tasks with the appropriate skills to an adequate level in near record time. Congratulations."

"Thank you, sir."

"You are about to take the next step toward your objective. It is a big step. Are you absolutely certain you still want to become a fighter pilot? The signs of war in Europe are near and far too clear. You have no requirement, no obligation, to be involved." Henry said the words he knew he had to say although in his heart he knew the country desperately needed more pilots with the skills, instincts and abilities of Pilot Candidate Brian Drummond regardless of their nationality.

"Yes, sir, more than ever."

"Yes, right. Well, then, I have here your orders posting you to Operational Training Unit Number Seven at RAF Hawarden. You are to report Monday morning. For now, you have a weekend pass to go anywhere you desire in Great Britain. Enjoy yourself, Mister Drummond. You have a fair amount of hard work ahead."

"Yes, sir," Brian responded taking the single sheet of paper from Wing Commander Henry. He started to leave Henry's office.

"I have not dismissed you as yet, Mister Drummond."

A bit embarrassed by his failure to remember a basic instruction snapped him back to attention facing his commander. Brian reprimanded himself to never make that mistake again.

"Despite your eagerness to leave Brize Norton, I had a few other things I wanted to say now that my official duties are complete." Henry rose from his chair to stand next to the nearly half of a foot taller young pilot. "At ease, if you will."

Brian relaxed his body taking the proper position of standing rest for members of the RAF. He was not quite sure whether he should face Jason

Henry or not, but he chose to remain official.

"I want you to know that I flew in a sister squadron with John Spencer and Malcolm Bainbridge. They were two of the more capable pilots in the RFC during the Great War. You've had an exceptional teacher in Malcolm Bainbridge. He holds a special place, a place of honor, for those of us who were over there. You may or may not know, Malcolm was the only American in the RFC. He was there, with us, because he thought he could make a difference and what's more he chose to remain with his mates even after the U.S. joined the war. He is a special person. And, as far as I have been told, you are the first and may very well be the only American in the RAF in the next great war."

Brian nodded his head slowly although he was nearly exploding with pride. His eyes began to water as he fought with his own emotions.

"As his protégé, it is an honor to know you," he said extending his hand to Brian.

Taking his proffered hand, Brian shook it firmly. "Thank you, sir."

"Now, one last word before I unleash you upon my beloved country and all those beautiful English women. Seriously," Wing Commander Henry paused to let the moment pass, "I am certain you will soon face the same dangers your mentor faced twenty-five years ago. I have just one piece of advice that I hope will help you survive the conflagration before us. Keep your head on a swivel. Keep searching the sky around you. Never let the bloody Hun sneak up on your tail. It will keep you alive, Brian."

"Yes, sir." Brian recognized a different set of words communicating the same advice Malcolm had given him so many times before. He took it as reinforcement of previous good advice. "I will do my best."

"I am sure you will. I look forward to hearing good things regarding your service, Mister Drummond. So with that, Godspeed. Now, you are dismissed, Mister Drummond. Get out of here and go have a good time. You have earned it."

"Thank you, sir," said Brian, then he turned to leave as he had been instructed.

Several of the other candidates as well as the instructors who flew with Brian gathered in the operations room to congratulate him. Some were jealous. Some were envious. Some were actually resentful. But, they were all respectful.

The words of accomplishment were appreciated, however short lived they might be. The appropriate words of thanks were also returned to the people he worked with over the very short two week period of his assignment at RAF Brize Norton. With the brief, unofficial and impromptu proceedings concluded, the others returned to their tasks for the day. Brian left the

Operations Building to the student barracks to gather his things.

"Brian, wait just a moment."

Flight Lieutenant Jeremy Morrison walked fast, nearly jogging, toward him with his flight kit in hand.

"Do you have any plans for your weekend pass?"

"Not really. I thought I would work my way up to RAF Hawarden, wherever that is?"

The short laugh communicated the 'oh you fool boy' thoughts of his former flight instructor. "First off, Hawarden is North Wales. Presumably, you do know where Wales is."

"Yes," responded Brian feeling slightly offended.

"It's just across the river from Liverpool. Nice place, actually. Good bunch of blokes. You'll learn a lot up there. Anyway, if you have no plans and I promise to get you pointed in the correct direction, what do you say to a tutorial weekend in London."

"With you?"

"Of course with me, you silly sod. I figured I could give you a proper orientation to London, I am sure quite unlike your friend, Group Captain Spencer."

Brian was not quite certain of the intent behind Morrison's offer. He had been a hard taskmaster, but Brian did feel the harshness was due more from caring than cruelty. "Sure."

"I have one more flight this afternoon which if I survive, we will drive to Oxford to catch a Great Western coach into town."

The intervening four hours were spent completing all the appropriate separation tasks associated with an individual's transfer from one military organization to another. While Brian did not possess the endemic knowledge common to other men his age who had grown up in, or around any military facility, organization or unit, he reminded himself continuously to take extra care with the procedures, the positions and the people. Malcolm's advice had so far proven to be precisely on the money although a few minor details had changed over the years. He would not allow himself anymore mistakes whether in the air or on the ground. He knew he had to try extra hard to do what was expected.

Several other candidates finished the day's training in time to say their good-byes. A few sought advice from the current phenomenon, now that he had completed eight weeks of entry-level training in two very intense weeks. The others wanted to perform at the same level although most lacked the aviation background. Some wanted instant success without the commitment to hard work.

With his institutional obligations complete, Brian waited at the candidate residence hall lobby. His thoughts wandered among the possibilities of what lay ahead, both immediately during a weekend in London with Jeremy Morrison to the next stage of his flight training and eventually to an operational squadron. All the Training Command officialdom to date stressed the importance of accomplishment on the ground and in the air. Only the chosen few could fly the leading edge fighter planes. So, do not get your hopes up young man, was the message sent and received so many times in the last few weeks. The words of diminished dreams served only one purpose for Pilot Candidate Brian Drummond. He simply had to redouble his concentration, focus and determination to achieve his goal.

"Are you ready for the time of your life, young Mister Drummond?"

"Yes, sir."

Jeremy stopped instantly and turned directly toward Brian. "Now, here is a direct order from a superior officer. As long as we are partying, I am just another bloke, and you will call me Jeremy or Jamie, if you prefer," he said sternly. "Is that clear?"

"Yes, s . . . ," Brian caught himself. "Sure, Jeremy."

"That's better. Now, let us see if we can set a new record from this humble place to the Oxford Rail Station. The rest of the journey will be controlled by the railway." Jeremy Morrison's diminutive, dark green, 1938 Morgan Roadster raced through RAF Brize Norton, past the protesting guards at the main gate and onto the Oxford Road before he finished his thought. "If we are lucky, we will be in the tender loving arms of a couple of dolly-bird friends of mine."

The fifteen miles to the station took just over twenty minutes although the trip ran four automobiles and two produce trucks off the road and startled no less than ten horses, some with riders and most pulling carts. The recklessness of the drive bothered Brian. He could not understand why Jeremy felt the need to take such risks. His discomfort over being drawn along to take the same risks without at least his consent occupied the majority of his thoughts.

The remainder of his thoughts considered the consequences of being with Jeremy's women friends. Being with an older woman, which they undoubtedly were, intrigued Brian from a curiosity perspective and a need to be accepted. The memory of Rebecca Seward existed quite vividly within his consciousness. His carnal thoughts regarding Miss Helen Riser of Windsor, Canada, or even his lustful images of Missus Mary Spencer made him feel a certain amount of guilt when dampened with the reality of his feelings for

Becky. He wanted Becky, here. It was not likely, but the feelings were still with him.

"If you are going to be this grim old fuddy-duddy all weekend, I will send you on your way now, Brian. You are not going to disturb my grand plans for satisfaction of my libidinous desires."

"I've just been thinking."

"Sometimes that is quite dangerous."

"What do you mean?"

"Well, like thinking about taking a shot or about what might happen if you jump enemy fighters. You can't think. You simply do."

"Why do you take so many unnecessary risks?" Brian blurted out.

"Well, well, don't we have quite the moralist." Jeremy conjured up a rather nasty false smile. "If you must know, this war is nearly upon us. Most of us thought we would be in it late last summer, but old Neville managed to buy us some more time. The other guys have more fighters, better fighters, more real training, not this pretend nonsense we do, and they are professionals at the game. We are more like a gentlemanly club that talks about past exploits instead of concentrating of what needs to be done in the next row. I do not expect to survive the coming storm, so I am determined to do as much hard living as fast as I can."

"That's a pretty black picture."

"It is a rather black future, I would say."

"Shouldn't we preserve every moment, then."

"I suppose so," Jeremy said, pausing to ponder. They both noted the Great Western Railway sign on the locomotive as it approached the station on time. Jeremy changed his mood. "Ah, here we are . . . our coach to pleasure."

"Well?" Brian pressed as they entered their assigned compartment, already occupied by an elderly, gray haired, wrinkled and slumped man.

A deep, hearty laugh carried on for several moments then tailed off. Jeremy looked at the old man. "Can you believe the naiveté of young people these days?" The old man only shook his head more in disbelief that he was asked such a question rather than disagreement. "Listen, Brian, there are only two types of people – savers and spenders. Savers save everything with the belief they may need something someday, and if they don't use it, they should give it to their children when they pass on. The spenders know they can't take anything with them, so they make the personal commitment to spend it all before they die. Well, I know I am a spender. I have a lot to spend and not much time to spend it. Therefore, I intend to live as fast as I can."

"What if you survive the war and live to a very old age?"

Once again, his question was greeted with boisterous laughter. "Spoken like a true saver. Simple, I am not going to survive this one. Maybe you will, but most of us will not," he answered with an unusual degree of solemnity. "Do you remember what you learned about formation flying?"

"Sure, but what does that have to do with it?"

"Everything. Listen to me. When you fly closer, the task becomes easier, but the risk goes up. It is quite like that in life. The harder you drive, the closer you get, the more risk you take, the easier life is to deal with, the richer the pleasures."

"That doesn't sound right."

"Maybe not, but it is Jeremy's philosophy of life. So, 'we who are about to die, salute you,' as the Roman gladiators used to say."

Jeremy sank into a melancholy which Brian tried several times to break without success. Brian resigned himself to watching the scenery pass by the rail car window. The remainder of the rail trip was taken in silence with both men considering the future, each with an entirely different view of the outcome and the consequences.

―

Friday, 7. July. 1939
Westminster, London, England

JEREMY Morrison did not return to his jovial character until they arrived at Paddington Station and entered the Underground.

Brian was given a thorough education in the subtlety of movement via the Underground rail system of London. The meaning of the symbols, the routing codes and the names of the important stations were discussed in detail, almost more detail than Brian could absorb at one time. They took several different lines with names like Circle, Central and Northern to arrive at Charing Cross Station.

The new arrival tasks – hotel rooms, freshen up, straighten the uniform – were completed quickly. Morrison was waiting for Brian in the lobby.

"Are you ready, my boy?"

"Sure," answered Brian without much enthusiasm. All other considerations aside, he probably would have been satisfied with a quick meal and a good night's rest.

Even in the middle evening hours, the streets of London were filled with people. The variety of attire, affluence, demeanor and animation still brought an abundance of wonderment to Brian Drummond. This evening, Brian's first night in the uniform of a RAF officer, carried additional thoughts and significance. For the first time, he began to feel the respect of a nation.

The nod of a head, the tip of a hat, the wink of an eye from total strangers gave him a new sense of worth. The people seemed to be genuinely appreciative, or proud, or something. Brian liked the feeling.

"Here we are," Jeremy announced dragging Brian quickly down a set of stairs into a dark, smoky night club.

Brian had not been able to keep track of where they were in London. The street signs as he was used to them in Wichita were non-existent. The mass of humanity, the buildings and the deepening night sky took away any sense of direction. They entered the club so fast, he did not notice the name.

The large room with sturdy pillars and several levels of tables forming a semi-circle around a stage or dance floor reverberated with laughter, conversation and pleasure. This seemed like a very happy place. An usher, owner, friend, Brian could not determine the relationship, led Jeremy and Brian to a table toward the side on the lowest level nearest the stage. Two incredibly attractive women were seated at the table with their backs toward the stage talking about something. One of the women had shoulder length, soft, silky blond hair that looked as smooth as a sheet. The other woman had curly brown hair much longer by popular standards.

The blond was the first to notice their approach. "Oh, Jamie," she squealed softly as she rose to wrap her arms around his neck, "my beautiful sugar cake. It's great to see you again."

Jeremy Morrison embraced the woman fully including a firm hand cupping one lobe of her buttocks. "It is great to feel you again, gorgeous." Jeremy turned to the other woman taking her right hand, kissed it slowly and looked up into her eyes. "It is good to see you as well, Anne."

"The honor is mine, Lord Morrison."

Brian heard the woman precisely. A combination of surprise, confusion and curiosity filled his thoughts as he caught Jeremy looking over his shoulder at him from his bent over position with a rather impish expression on his face. Was this woman joking, or simply stating a well-known, to all except Brian, fact of title.

Without saying anything to Brian, he returned his gaze to Anne. Brian could barely hear him saying to Anne, "My dear, you know better. Please do not call me that again. It is Jeremy or Jamie – one or the other." A quick smile broke the tension.

"As you wish. My apologies."

In a more normal voice and now standing fully erect, Jeremy said, "May I introduce our young savior, the soon to be pride of His Majesty's Royal Air Force, Mister Brian Drummond." Each woman extended her hand which

Brian grasped firmly to shake. "Brian, this is Miss Virginia North," he said motioning toward the blond. "And, this is Miss Anne Booth."

"Nice to meet you both."

"Now, Anne, doesn't he have the most delightful accent?" Virginia queried in a soft, elegant English voice.

"He certainly does." Anne's coy smile and lowered head portrayed a sense of shyness. The impression offered a brilliant contrast to the seductive, deep blue dress cut low to accentuate the cleavage between her ample breasts. As Brian's gaze reflex alternated between her bosom and her sapphire blue eyes, he gradually became aware of the change in her smile like a wolf confident she had captured the hare with her eyes.

The spell was broken by a hard slap on the back. "Not so fast, my boy. In time." Brian felt the hot wave of embarrassment singe his flesh. "In time. We have a show to watch with the pleasures of these ladies."

"I'm terribly sorry," Brian offered to Anne with the searing heat of his embarrassment still glowing in his cheeks.

"You are so sweet," Anne responded gently touching his cheek. "He is genuinely embarrassed." She leaned over near Brian's ear and whispered. "It is quite all right, Brian. It happens all the time. I am quite used to it. And, I am glad you appreciate them."

"Ladies and gentlemen," the master of ceremonies began, "the Harlong Club proudly presents for your enjoyment, the Pleasures of the Flesh."

The characteristic rumbling beat of the big band swing music began the show. A beautiful dancer dressed in a loose and flowing gown floated across the stage moving exquisitely to every note the band played. As she whirled, slid, turned and jumped, Brian could not mistake Anne's hand resting on his thigh. The dancer began to remove parts of clothing as the music changed tempo. As more flesh was exposed and Brian's attention was split, Anne began stroking and edging her way up his leg. The touch of her hand was making his skin crawl with excitement. The rest of his body was responding to the visual as well as tactile stimuli. His heart was pounding as the dancer released her gorgeous breasts from their confinement.

An element of shock mixed with fascination. The only breasts he had ever seen were Becky's but that was in the privacy of their own intimacy. These breasts were in public for all to see. At that moment, Anne's hand reached its objective to discover the extent of Brian's excitement.

Brian's emotions, feelings, propriety and sense of right burst under the overload of stimulation. Without a word, a sign or warning, he leapt from his chair and virtually ran from the club. Out the door, he began to run down the street not really knowing where he was going.

"WHAT happened?" asked Jeremy without any tone of accusation and looking at Anne.

"I was just touching him. He had an erection."

"You might be moving a bit too fast for an eighteen-year-old, Anne dear, and just so."

"He is only eighteen?"

"That is correct."

"He is too young to be in the RAF."

"I am afraid not, my sweet. In fact, we have some who are younger than him."

"I didn't know."

"Well, now you do. You might also want to know, he has only been in England for about three weeks or so, he is from the middle of America so I am sure he has not been exposed to sophisticated dollies like you."

"Thank you, Jeremy. I will do better."

"I hope so. I have paid you handsomely, in advance I might add. I am only asking you to make this a weekend he will never forget, and it appears we have lost one night of it. I won't bore you with the speech about him being a young pilot facing certain death in the coming war. He is a good kid, and I want him to have fun. I want him to feel the best London has to offer."

"I will take care of it. Where is his hotel room?"

"I would suggest you leave him alone tonight. We will try again tomorrow."

Saturday, 8.July.1939
Westminster, London, England

BRIAN spent a good portion of the night and most the morning considering what happened the previous night and grappling with the explanation he would offer to ease the continuing flood of embarrassment.. He wanted to be accepted, to be a part of this new world he sought, to be sophisticated and worldly. He was simply ill-prepared to deal with so many emotionally powerful events. Growing up in Wichita, Kansas, had certainly not prepared him for what he experienced last night. The undeniable attraction to a beautiful, even if older, woman brought with it a sense of guilt. The inevitable question between separated lovers added to the turmoil. Was Becky being faithful to him? Was she under the same kind of pressure to transgress the same temptation?

Part of his conscience told him to remain true to Becky. He loved her. She meant a great deal to him, more than he realized when he left the United States. Being apart brought home the feelings he possessed toward her.

The conflict came in the form of curiosity. What would it be like to be with a beautiful and apparently experienced woman? What wonders lay just around the corner? There was a real and growing part of him that wanted to find the answers to the questions. The attractions of the female form were irrepressible. Jeremy Morrison obviously enjoyed himself around women. He was probably very experienced, as well. Maybe most RAF pilots were far more experienced than he regarding women and sex. Maybe that was part of being a pilot. Maybe that was expected of RAF pilots. Malcolm had never told him anything about this part of it. Was it actually the same for him and he just had not told him, or what?

The telephone in his room rang. "Are you ready to get something to eat?" asked Jeremy.

"Sure," Brian answered although he had already eaten a late breakfast. He wanted to know what kind of damage had been done by his rapid departure.

"I will meet you in the lobby in ten minutes."

"OK." Brian rehearsed one last time the reason he would offer for his actions last night.

They walked into the hotel dining room, sat down at an empty table along the wall and ordered up a light meal of sandwiches.

"You missed a good show and a great time last night."

"I was just tired."

"I know what you mean," said Jeremy knowing precisely that his departure had nothing to do with being tried.

They spent lunch talking about flying. Questions were asked and answered about RAF Hawarden, operational procedures, Spitfires and Hurricanes, neither of which Jeremy had flown yet, and RAF customs. Several attempts to change the subject of conversation to any direction other than professional were resoundingly rebuffed by the young American pilot.

The discussion served another purpose. The worries, concerns and apprehension from last night's embarrassment wafted off into the distance. Jeremy also had a plan to help young Brian forget about it.

"I never realized until just now how consuming your objective is. There aren't many men who are so eager to face death in the confines of a small cockpit racing around the sky. I'm not sure if it's the foolishness of naive youth, or the commitment of a determined man. I suspect the latter. If so, I also suspect Great Britain will soon be immensely grateful for your commitment."

Brian felt a twinge of flush with Jeremy's kind words. "Thank you and I'm not naive. Malcolm made painfully sure I understood the ugliness of warfare. He wanted me to know what I was getting into. I've wanted this for some time, and I think it is a commitment."

"Then, glory hallelujah. Now, let's see if we can find some trouble."

"One last question."

Jeremy relaxed back into his chair. "What is it?"

"Anne called you, Lord Morrison. Does that mean what I think it does?"

"It depends on what you think."

"Are you part of royalty?"

Jeremy was obviously not pleased with the topic. "I will do this once and only once. Agreed?"

"Do what?"

"Answer your questions about this topic."

"Agreed."

Jeremy paused for a moment to look around at the other empty tables and confirmed with his wristwatch the mid-afternoon hour. "First off, let me say I do not think of myself as royalty. I think of myself like any other citizen in the service of his country." He paused again. "I am the third son of the 7th Duke of Cottington. My older brother is the 8th Duke, so I suppose the peerage would consider me nobility. I let my family do what they want, and I am doing what I want, much against their objections I might add."

"Do you have a castle?"

"My family does. It actually belongs to my brother, the 8th Duke."

"Am I supposed to call you Lord, or your highness, or something?"

"If you do, I will clot you square in the mouth. Anne knew better. She made a mistake."

"OK." Brian was not sure if he was supposed to feel differently toward Jeremy or not. Jeremy Morrison made his desires perfectly clear. So, Brian concluded, to treat him as just another friend and fellow pilot. "Then, different subject. Is Virginia your girlfriend?"

"In a manner of speaking, I think you could say that."

"Were you friends with Anne?"

"Not really. She is Virginia's friend. I thought they would be good company for us to have fun with this weekend."

"They are both very attractive women."

"Yes, indeed."

Jeremy debated with himself whether he should ask the next question, not wanting to spoil Brian's good mood. The plan, however, required it. "Do you want to see Anne again?"

"I don't know if I should."

A strong laugh punctuated the conversation. The waiter came up to the table to ask if there would be anything else suggesting with his tone that they should move on. Jeremy checked his watch once again. It was nearly tea time. "We will have some tea and biscuits, please." The waiter was clearly not impressed but did as he was asked. "Are you a man?"

"Yes."

"Maybe I should not ask such a personal question, but do you like women?"

"Yes!"

"Well, then, there you have it. Of course, you should. What on earth would make you think you shouldn't?"

"I have a girlfriend back home. Her name is Rebecca Seward. I call her, Becky."

The waiter returned with a tray containing a pot of tea, two cups, milk, sugar and a plate of assorted cookies that looked exceptionally good. Jeremy did not even acknowledge the delivery because he was intent upon responding to Brian.

"Good English name. She must be a very nice girl, then. Anyway, she is there, you are here. So, what seems to be the problem?"

"I want to marry her and have a family of my own."

"So?" Jeremy asked as he poured the tea.

Brian followed his lead. "I need to remain faithful."

"Dear boy, you have so much to learn, and you are lucky to have found me. I am just the person to teach you the ways of the world. Now, listen closely. You can remain faithful to your future wife and mother of your children without denying yourself the pleasures of the flesh."

"I don't understand."

"I am sure. Well, then, let me see if I can explain." Jeremy paused as if he were contemplating how to put into words a difficult topic. "Your love for your wife should be thought of like a river that just keeps flowing year after year. A tryst of pleasure is simply that . . . temporary, passing, to be enjoyed for the moment. You are not married. You are not even engaged or betrothed, yet. You are just a young bull out smelling the flowers."

"What if you develop feelings for the woman?"

"Then, so be it."

Brian sipped his tea which had a rich, creamy taste and enjoyed a cookie as he considered what Jeremy was saying. The logic seemed reasonable, but there was still a part of him that did not agree. At this point, the curiosity of pleasure weighed in the balance.

"Maybe it wouldn't hurt anything."

"That's the spirit. With that weighty dilemma resolved, shall we move on to our next adventure?"

"Which is?"

"Shepherd's."

"Which is?"

"Good that you should ask, my boy. Shepherd's is a drinking establishment quite often frequented by our brethren."

"Pilots?"

"Oh, my, my, not just pilots, dear boy . . . fighter pilots."

The distance to the renowned pub was not particularly far in a central section of London called, Mayfair. The taxi ride took longer than it should have due to the peculiar mixture of vehicular and pedestrian traffic on the sometimes narrow streets. The building was a familiar Tudor structure with dark wood beams and white plaster walls. The large sign displayed the name underneath the painting of a rolling Spitfire over the English countryside. The place was larger than it appeared from the exterior. A long, traditional bar complete with brass footrest dominated the room. Tables and booths occupied the remainder of the space. The smoke obscured his vision and burned his eyes. The smell of burnt tobacco and stale beer overwhelmed him. Quite a number of men dressed in the characteristic and distinctive blue, gabardine uniform of the Royal Air Force officer gathered in small groups. In the early evening hour only several handfuls of women blended with the officers. Brightly colored as well as various printed dresses added to the jovial atmosphere that greeted the two men.

Several of the officers recognized Jeremy Morrison as a long lost comrade returning from some distant land. Full pints of dark English bitter were thrust into their hands resulting in some spillage. Jeremy dutifully introduced Brian to his colleagues although no one seemed to really notice he was there. While Brian thought of himself as a fighter pilot, one of them, he reminded himself he did not have the wings of a pilot, yet. Therefore, none of the patrons looked at Brian as anything other than a friend of Jeremy Morrison's and as introduced – an 'almost.'

The recollections, stories, jokes and exaggerations occupied Brian's attention quite well. The words, laughter and good-time talk covered only two subjects, aviation and women, far more the former than the latter. Names like Deere, Malan, Bader and Stanford-Tuck repeatedly popped up in the conversations. They were spoken in a peculiar manner of respect verging on reverence. These men possessed unique attributes, skills, traits, instincts or characteristics that set them apart from other very accomplished men. Brian remembered the names, remembered the stories, and vowed to learn from the successes and failures of others. He intended to be the consummate student and perform to the maximum of his capability. Despite the accolades received to date, Brian knew he had a lot to learn from the men around him. Trying not to be overly impressed or awestruck, Brian also conjured up the voice of Malcolm Bainbridge reminding him of all those valuable lessons.

Much to Brian's amazement, no one seemed to be paying for the drinks. Before a pint glass with its etched crown could be fully emptied, a fresh, full glass replaced it. The laughter and camaraderie were contagious.

Several hours must have passed before Brian realized more officers and many more women had arrived filling the place up. The words and mood began a noticeable shift as more attention moved to the women. Some were obviously couples. Most seemed to be single.

"Careful, my boy," Jeremy said into Brian's ear. "A good many of these voluptuous, attractive, young dolly-birds are actually vicious, merciless predators hunting the unsuspecting young pilot who might fall under their spell."

"Oh, sure."

"Very well, then. Beware. Forewarned is forearmed," Jeremy said with the lightness and humor it deserved and returned to the conversation with his friends.

Brian widened his awareness to take in the other groups of pilots, some with women, most without. A larger number of the women appeared to be moving about among the sea of blue uniforms. The image struck Brian as a visualization of Jeremy's comment. He laughed more to himself at the observation, but also did not fail to note the point. He was not quite sure what he would do with the advice.

Near the entryway, the bright golden hair of a tall woman emerged above the heads of the patrons. The sea seemed to part as she moved into the room. Virginia North. She was looking for someone probably Jeremy. Before Brian could move to alert Jeremy, he wanted to know the answer to his own question.

There she was walking slightly behind Virginia. Anne Booth's delicate, subtle attractiveness stood in clear contrast to the flashy, luminescent beauty of her friend. She was wearing a light, flowery printed dress which was predominately blue and green, buttoned down the front from her neck to her knees. Brian noticed immediately the dress hid her bosom.

Brian grabbed Jeremy's elbow. "Virginia and Anne are here."

"Great," he said. It took him a fraction of a second to locate his girlfriend. "Virgin," he shouted above the din.

There was no mistaking the word Jeremy used. Brian looked to his friend with surprise and incredulity. The impact on many others in the room was equally as dramatic. The only one that did not seem to be affected was Virginia who smiled and waved. They embraced and kissed deeply which drew a resounding chorus of oooh's and aaah's along with other more pointed chants and encouragements.

"It's good to see you again, Anne," Brian nearly shouted over the ruckus created by Virginia and Jeremy.

"I am fortunate to have been blessed with a second opportunity to meet you, Brian," she said leaning so close to his ear he could feel her warmth.

Brian offered Anne a drink which she accepted. He made several fumbling attempts at conversation but found it quite difficult to talk about anything other than aviation. Any subject he thought of seemed inappropriate for a chat with a sophisticated woman.

Anne let him wander around a bit. Although he was quite young, younger than any man she had been with since she was the same age, a strange, magnetic attraction possessed her. She needed to guard her emotions. This was business. She had been paid in advance, handsomely paid, to make this young pilot feel good. His naiveté brought a pureness which she had not experienced in several years. Yes indeed, she reminded herself repeatedly, she needed to maintain her guard or this man could get to her heart.

"I'm rambling, as my mother sometimes tells me," Brian finally admitted aloud as he recognized his nervousness. He was talking, but not saying anything.

"It is hard to talk with all this going on around us. Do you want to go outside for a walk, so we can talk?"

"Sure, that would be great." He thought about telling Jeremy he was leaving but decided not to. Jeremy was not his father or his older brother. He had no requirement to tell anyone what he was doing with his spare time.

The street was not much quieter since they traded the cacophony of human voices with the sounds of the city street. They walked side by side for quite some way. After several blocks, Anne took Brian's arm drawing them ever so slightly closer. A large park with its large trees, wide, sparsely lit walkways became the majority of their pedestrian environment. Anne provided the subtle guidance in terms of direction without really saying where they were going. Brian did not much care. He simply enjoyed being close to such a beautiful mature woman. They talked about many subjects from the sights of London to the character of the Great Plains of Kansas. Anne had never visited any country beyond Europe. There was a genuine fascination with America. She asked as many questions as Brian did and tried to stay at his level of sophistication, a trait she learned well from Virginia.

"Would you like to get something to eat, Brian?"

"Sure." He was feeling signs of hunger.

"Would you mind having something at my flat?"

Brian thought nothing was unusual or odd about the offer. After all, they were adults, now weren't they. "That would be fine."

"Great. I live just a short distance from here. We can walk if you don't mind?"

"No problem for me."

The conversation continued along the same lines as they moved among the varying mass of citizens of the great city. The narrow, white stone building with a strong, black, wrought iron railing, five stone steps and heavy, black door with polished brass knocker was two or three miles from Shepherd's, as best Brian could tell.

"Here we are."

The interior was light and airy with a pleasant mixture of light colored woods, modest furnishings and a delightful, aromatic, floral fragrance. Brian was truly impressed.

"This is a very nice place."

"I like it. It's simple, functional and warm. That's about all I need."

"It must be very expensive, too."

Anne knew this was not a topic of discussion she wanted to get into. It led directly to her profession which would not contribute to the accomplishment of her tasking. "My parent's flat, actually. They let me use it when they are away."

"Where are they?"

This too was an area Anne did not want to be wandering around in. "Let's not talk about my parents and real estate, shall we?"

"OK." Brian considered the sensitivity of her request, but he also could not think of anything else to talk about other than aviation, and he knew that was not the right subject either.

Quiet filled the room until Anne could switch on a phonograph for some contemporary, big band, music.

"Make yourself comfortable. May I take your jacket?"

Brian quickly unbuckled his waist belt, unbuttoned his tunic and handed it as well as his hat over to Anne. He was becoming accustomed to the shirt and tie, but the RAF uniform tunic and hat still took some adjustment and felt very good when they were removed. Anne hung the jacket carefully and neatly in the entryway coat closet.

"I do not have much in the house," said Anne not entirely truthful. "Would you mind a simple sandwich and some cheese 'n biscuits? I also have some wine, or we can have tea."

"That would be just fine, and I'll have some wine," he answered thinking the wine would make him seem older than tea. He was still occasionally reminded of his age although it was most often a self-imposed reminder.

The warmth of words helped Anne get to the point she wanted to be. She had avoided discussing the previous night's incident and noted quite well Brian had done the same. She also knew she needed to start over and try to erase the first impression.

"Brian, if you will permit me," she hesitated in an attempt to gain his consent. Brian did not know quite what to expect, so he nodded his head with a curious expression in his eyes. "I must apologize for my behavior last night."

A slight flush of embarrassment returned to his cheeks. "Forget it."

"I would like to, but I offended you and embarrassed you in front of your friend. For that, I am deeply embarrassed myself and very much regret placing you in that position. Sometimes I assume too much. I suppose I have been biased by pilots like Jeremy. They all seem so fun loving and uninhibited. I did not realize you felt differently. Please accept my apology."

"No apology necessary," Brian responded feeling more embarrassed by the fact Anne thought he was different from the other pilots, that he might not be a pilot, a fighter pilot. He wanted to be a fighter pilot. He simply had not considered the social or personal side of being a pilot, or the image of what a fighter pilot should or might be. He needed an excuse to put last night behind him. "I have a girlfriend back home." He could not think of how else to say it. "I was pretty tired, too."

"Then, can we start over?"

"Absolutely."

"Splendid. Then, would you let me know when you want to go back to your hotel, and I will make sure you get there safely."

"You don't need to do that but agreed."

"Excellent. Then..." Anne stood and Brian followed. "Good evening. My name is Anne Booth," she said with the softest, sweetest voice he had ever heard as she extended her hand.

Brian took her hand gently. He felt her warmth that produced strange electricity within him. "Good evening, ma'am. My name is, Brian Drummond."

"I prefer, Anne, if you please. So, you are in the RAF?"

"Yes."

"You are not a pilot. What do you do?"

"I'm a candidate pilot. I'm in training."

"Very good."

They both paused not quite knowing what to say next. Anne knew exactly what she wanted to do which was exactly what Brian wanted. Anne hesitated trying to judge if she might be moving too fast. The reluctance within Brian came from a different source although it produced the same result.

"Would you like to dance?" asked Brian finally since it seemed they were pretending to be at some public meeting place.

"Certainly."

Even the light touch of a proper dance position brought very distinct responses within each of them. Anne recognized the phenomena precisely and was pleased with the undeniable attraction. She knew at that instant this was no longer business despite what Jeremy Morrison had done to arrange their meeting. For the younger man, a different recognition produced similar responses. One quick thought of Rebecca Seward was his last for this evening.

The dance became slower and closer as they ignored the tempo of the music and moved deeper into the forest of intimate affairs. The soft curves of her body ignited an inextinguishable fire which Anne easily detected without overt effort. An inner smile gave the last fraction of confidence she was looking for.

Without another word, Anne took Brian's hand and led him upstairs. For his part, Brian felt helpless, powerless as well as excited. The urge to quicken the pace and take more aggressive steps built up a tremendous pressure in his young, virile body. Anne took full control of events taking deliberate, careful actions with each button, lace, tie and zipper. He stood before her completely naked and in full glory as he watched her disrobe at a desperately slow rate. He wanted her more than anything he had ever experienced before. The world beyond her ceased to exist.

Her hand ever so slowly lifted his hand to her breast. Brian withdrew his hand sharply as if he touched a hot kettle. The sight of Anne standing before him without the slightest flicker of embarrassment added to Brian's curiosity, confusion, desire and failure of his resistance.

"Brian, it is quite all right. Flesh is to be enjoyed, savored, rejoiced and appreciated." She gently grasped his hand again. "Please, Brian, enjoy my flesh." The roundness of her full breast shattered the last vestiges of any pretense of resistance. "Absorb the curves, the smoothness, the texture, and the hardness of my nipple." Once satisfied his passion had taken hold, Anne moved her attention to his lean, hard, anticipating body.

Anne Booth used her experience in matters of the flesh to heighten his pleasure. The energy between them was enormous, more than either of them had ever felt which was more significant for Anne. The touching, kissing, fondling, caressing between a succession of intimacies brought an even greater appreciation. It was more than half way through the shortened night of the summer months when Brian finally succumbed to exhaustion and lapsed into a deep sleep.

A<small>NNE</small> lay next to him with her hand on his chest feeling the rhythmic breathing so characteristic of deep sleep. She relished the satisfaction she sought, but could not deny the satisfaction she wanted. Thoughts wandered into the future. The energy between them yielded its own pleasure. He was so young, inexperienced and fresh. The opportunity to teach him the techniques of a great lover, to channel his energy toward her pleasure as well as his own, to make him what she knew in her heart he could be, had a cloud darkening the light. How could she move away from her past? Was it possible? How could she do it? Why would she do it? My God, she told herself, he is a young air force pilot with a war looming ahead of them. What possible future could he have? . . . could they have?

Chapter 7

> For unto whomsoever much is given,
> of him shall be much required:
> and to whom men have committed much,
> of him they will ask the more.
> -- Luke 12:48

Monday, 10.July.1939
RAF Hawarden
Broughton, Clwyd, Wales

THE administrative building to which Pilot Candidate Brian Drummond had been directed was virtually identical to the one at RAF Brize Norton, a plain clapboard structure with simple, shallow pitch roof, asphalt shingles, and all painted standard, military, medium green. The warm, summer sun added some wetness under his uniform tunic. It was the warmest day he had experienced in Great Britain, so far, but it was quite moderate compared to summers on the Great Plains of Kansas. The rich, green, Welsh mountains to the West that dominated the horizon, and the people, the buildings, the atmosphere of the facility were consistent with his limited experience. The ancient, walled, Roman city of Chester offered additional stimulation to his curiosity.

"Good morning, sir," said the clerk at the counter, a female RAF aircraftman, the first that Brian Drummond had seen in uniform.

"Good morning, miss. I'm here to check into Operational Training Unit Number Seven."

"Yes sir. Your name?"

"Pilot Candidate Brian Drummond."

"Yes sir. I have your papers right here." She shuffled through a one inch stack of identical papers to retrieve the proper item. "You are assigned to D Flight for advanced flying instruction. Here is a list of check-in procedures for you. You should report to D Flight Operations, just down the road, third building on the left, this afternoon for your flying assignment. If you have no questions, we'll get you started on your training." Meaning, he could process himself through checklist.

Brian looked at the list of items covering the events of the first few days, operations procedures, the name of his principal flight instructor, quartering, medical. They were essentially identical to those he had seen at RAF Brize Norton.

As he proceeded through the day's tasks, his thoughts were not of the actions that were purely administrative in nature. It would be another day or two before he could fly again, he told himself. The majority of his conscious thought was of Anne Booth and the events of the weekend, all the feelings, all the meaning, all the concerns, all the possibilities associated with that night and the incredible morning after. He still felt the exhilaration of new discovery, and unfortunately in the light of day, it was tainted with a stain of guilt. His thoughts varied between Becky, whom he was sure he still loved very much, and Anne Booth who had an attraction, a very powerful seductive attraction, for him.

Check-in procedures for the day went quite easily. He seemed to be familiar enough, although he had only done it once before. He received his flight assignment. His flight kit had been checked out by the technicians. He had received a couple of manuals, much to his disappointment; the most significant booklet was the Pilot's Notes for the Gloster Gladiator.

One flight of Spitfires, six aircraft, as well as one flight of Hurricanes, were quite noticeable on the flight line as he moved about the airfield. He had been told by one of the mechanics the Spitfires belonged to A Flight, the Hurricanes to B Flight. He wondered why he had not been assigned to A Flight, but knew that he was not in a position to argue or request. He still had much to prove, much to learn, and many skills yet to polish.

In the common area of the residence barracks, several fellow students gathered to discuss the day's events. Each of the other candidates in turn welcomed Brian as the new guy. The names came so fast he could not remember them. Each of them had completed different stages of advanced flight training. It was not until another jovial pilot candidate entered the room that he met his first companion at the new assignment.

"Good afternoon, Jonathan Kensington here," the young man announced. About six inches shorter than Brian, light brown, almost blond, hair, blue eyes, he looked very much like the German posters that he had seen in various places, describing the Nazi view of the Aryan race.

"Good afternoon, my name's Brian Drummond."

"Aye, yank eh?"

"That's right," Brian responded wondering what was coming next.

"Well, it's good to meet you. Checking in as well I see."

"Yes I am. It's my first full day."

"Mine as well," Jonathan said. "This barracks belongs to D Flight."

"Where did you get your primary training?"

"Brize Norton"

"Excellent!"

"Where did you get your primary training?" Brian asked.

"Bristol, actually."

"Where's that?"

"Well, south of here, south, just across the river from South part of Wales."

"I haven't been down there," Brian added.

"What part of America are you from?"

"Kansas."

"Ah, the Great Plains, the bread basket of America, the central heart of the Unites States. I have never been to America, but I would very much like to."

"Well maybe someday I'll be able to show you my country."

"That would be great."

"Would you like to get a beer?"

Brian wondered whether he should drink at all given the possibility of an initial flight evaluation like he had with Wing Commander Henry. If the process was the same, he needed all his faculties and concentration. The amicable nature of Jonathan and the affinity that seemed to exist instantly between them convinced him to take the risk. "Sounds good to me."

The two pilot candidates asked the other students, the other candidates, if they wanted to join them for a beer. None of them accepted. They seemed to be preoccupied with their studies for impending flights. Brian and Jonathan proceeded down the roadway back towards the officer's mess building and into the bar for the officers. It was a dark, wood paneled, area in a different room of the building, with numerous plaques, trophies, pennants and other memorabilia from the operational training unit attached to the walls. Several pilots, most of them appeared to be instructors, were already in the bar drinking several beers talking about one thing or another. All of them seemed to ignore the two pilot candidates. Brian followed Jonathan's lead. Jonathan seemed to know what he was doing, so he saw no reason why he should not follow his direction. He asked the barkeep for two pints of bitters, which they accepted. The two young fledgling pilots sat down at a quiet corner table out of the way, trying not to interfere or mix with the instructors since they would determine their fate. They talked a great deal about what they had done, where they had come from, how they had grown up. Jonathan was absolutely fascinated by Brian's flying experience, asked many questions about Malcolm Bainbridge, Malcolm's experience during the Great War, and what it was like to fly fighters. He very much wanted to fly fighters. Jonathan was the first in his family to go into the military. His family had been merchants for generations. None of them had really served in the military before. Jonathan felt a duty far beyond the rest of his family and despite the objections from his parents, he had joined the RAF. He had seen aircraft flying around the countryside and in a very similar way to Brian, knew that he had the same calling. It was quite apparent to Jonathan that he did not have the flight experience and the flight time that Brian had. The desire was certainly as great and the need to satisfy that desire was equally as strong.

Although Jonathan never said the words directly, Brian detected the distinct signs of a man, not much older than himself, trying to break free of something, maybe his father, maybe his family, or maybe just a spirit from his ancestral past. Jonathan was definitely a man with something to prove.

Brian felt himself drawn to Jonathan's family and the environment, the type of livelihood that they had, this was the first real contemporary of his that he had been able to sit down and talk to on a more personal basis about their living experiences. Jonathan was quite a jovial, open, candid person who freely shared his experiences and his feelings. The friendship between the two young pilots was borne in those conversations, in those discussions, and they knew that they shared a common desire, a common goal and a common objective, while their heritage and their past were quite dissimilar, from different cultures, from different backgrounds, from different classes for that matter. They shared the same desire to be a fighter pilot. By the time the two young men realized the time it was well into the evening. They both suspected they would have a difficult day tomorrow. They, as student pilots, would meet their principal instructors and begin their introduction in the Gloster Gladiator, the aircraft they would soon learn to fly and use for the intricacies of fighter tactics and operations. Both men retired for the evening with very much a content feeling. Brian knew he had found a kindred spirit in Jonathan, but his thoughts soon led back to Anne and the events of two nights previous.

Tuesday, 11.July.1939
RAF Hawarden
Broughton, Clwyd, Wales

MORNING call came early for both Brian and Jonathan. They ate breakfast, gathered up their flight kit and headed down for their appointed time at the D FLIGHT OPERATIONS building at 09:00 as their instructions indicated. Each met their own instructor who then took them aside into separate rooms and began going through a set of procedures and processes for the first period lessons. He covered the basic course of instruction, the identification of those elements that were critical, and the criteria for passage from one stage to the next. After his instructor, Flying Officer George Esling, had finished with what Brian had surmised was the formal portion of his orientation and instruction, he decided to add a more personal statement.

"Mister Drummond, I think I owe you some candid observations that I hope will assist you in successfully completing this course. We have all been told about your reputation at Brize Norton and the impression you have left with your primary flight instructors, including Wing Commander Henry, whom we all hold in very high regard, but I want you to know one thing, one very specific thing. There is only one reason. There is only one way. There is only one process by which you are going to complete this course successfully, and that is by performing, doing what you are told, what you are asked, what is expected and performing to the best of your ability. It does not matter how much you have flown before. It does not matter what you have flown before. It does not matter who you know. It does not matter what you think you can do. All that matters here is what your actions demonstrate. What you can show to me and the other instructors of your capability. So you need to forget about everything that has happened before. You need to pay attention. You need to listen, and you need to do exactly what you are told. Is that clear?"

"Yes, sir." Brian staggered within himself from the weight of his words, then hunkered down to deal with yet one more obstacle placed before him.

"Excellent. Well then, if we have that straight, let us get on with the business at hand. Grab your kit. Let us go to the aircraft. I will give you a good preflight, a good description of the aircraft, and then we will fly." Flying Officer George Esling did not wait for a response stepping out smartly toward the Gladiator line.

Brian was quite surprised to be faced with the reality that he was going to be flying sooner than he had expected. He regretted not reviewing his Pilot's Notes for the aircraft the night prior. He did not even know how to start the aircraft and hoped that he would find some familiar switches, although he could tell from a distance the Gloster Gladiator was certainly much bigger and more powerful than the Tiger Moth he had flown. But, there was no turning back now. Esling expected a certain capability and he very much regretted his reputation and the fact that he had not studied the Pilot's Notes prior to starting the day's events. He walked with his instructor down the flight line to the area that belonged to D Flight. As they approached the aircraft, the differences became very obvious. The Gloster Gladiator was by far the biggest biplane he had ever been close to. A big radial engine, an enormous, two-blade, wooden, fixed-pitch propeller and an enclosed, single place cockpit with a set of steps you needed to get into the cockpit. He could not even see into it, it was so high up. It was a big aircraft, and it had guns. More importantly, he would solo on his first flight since the Gladiator was a single seat, operational fighter, not a trainer.

His instructor walked around each and every element on the aircraft, pointing out all the important pieces – landing gear, control surfaces, brakes, engine. It was certainly the largest radial engine that he had ever seen. Larger than the modified Mystery S he had flown in St. Louis not too many months ago. The wings, ailerons, struts, support wires, tail, fuselage, all of it. He covered every inch of the aircraft and described the significant pieces for each and every portion. He went through a very thorough description for the aircraft and it's capabilities.

Then, Esling asked the dreaded question. "Did you study your Pilot's Notes, Mister Drummond?"

"No, sir."

"Well, that is most unfortunate and quite unprofessional, but I guess we will get a good glimpse of how good you really are. Hop in the machine, I will talk you through the start procedures. If you have any questions, I dare say, you should ask them. I will be flying aircraft G right next to you on your wing."

Brian's heart rate increased. He knew the pressure was on. Here was an aircraft, the size of which he had never flown before, had never experienced. There was a fair amount of apprehension in his heart as he donned his flight kit including his Mae West – the personal flotation vest so named for the well-endowed actress – parachute and gloves, climbed into the cockpit, sat down in the seat and strapped himself in.

Flying Officer George Esling stood up beside him, quickly pointed out the start switch, the magnetos, throttle, fuel mixture. They were all controls he was familiar with, but the size of this machine was still quite intimidating. He tried to absorb as much as he could of the preflight, prestart and start procedures. It seemed pretty straightforward. He thought he could handle it. Flying Officer George Esling asked him one last time if he had any questions. Brian shook his head.

Esling concluded, "I will call for taxi. We will line up into the wind. I will take a position behind your left wing. I want you to look to me, when I nod my head go ahead and take off. I will be following on your wing and talking through the procedures. Now remember Mister Drummond, this aircraft is probably bigger and more powerful than anything you have flown before, so you will have to be keen on the rudder to ensure that you keep the aircraft straight on your takeoff run."

Brian wanted to acknowledge verbally but could not find the words. He simply nodded his head in agreement. As his instructor worked his way over to his aircraft and strapped into the cockpit, Brian quickly went through the cockpit, scanning the switches and gauges, to familiarize himself with what he thought Esling told about starting. It was fairly straight forward. He expected to have no problem. He saw the engine start on the aircraft adjacent to him. He figured that was an obvious sign that he was expected to start. He signaled to his crew chief standing beside the right wing by holding up two fingers and making a quick circle that he was going to start.

He opened the throttle slightly as he would expect. He pushed the starter, the engine cranked two or three turns. He switched the magnetos to BOTH. The engine kicked and fired belching a large, blue-gray cloud heavy with the pungent odor of partially burnt gasoline. A big, loud, heavy sound rumbled through the aircraft as the engine turned over. The blade of the propeller seemed to move slowly as it turned in its enormous arc. He could feel the power in the aircraft, even though the engine was still at idle. He checked all of his gauges. Oil pressure was coming up. The engine seemed to be running normally. He looked over to his instructor. Esling signaled him to taxi out. Brian watched and followed his instructor, advanced the throttle and taxied the aircraft out of the line. The aircraft took a fair amount of rudder to hold the nose straight. They moved down the edge of the grass landing area. Brian followed, took up position as he was told and advanced slightly ahead of his instructor's aircraft. He kept looking over his shoulder. Esling kept pointing at his ears. He was not quite sure what he was referring to and shook his head that he did not understand. The response was a clear indication that he was not doing well. Flying Officer Esling had a disgusted expression on his face, simply shook his head and signaled Brian to takeoff. Brian looked into the cockpit, checked all the gauges, made sure everything looked about right and slowly advanced the throttle to make sure the engine was going to increase RPM before he released the brakes. The aircraft started to move and gain speed, he advanced the throttle as Malcolm Bainbridge had told him so many months ago on the Mystery S. The aircraft rumbled down the grass strip, bouncing, shaking and shuttering. The tail rose into the air which he held level to make sure that it did not over rotate and dig the propeller into the turf. The aircraft handled fairly easily and soon was in the air and flying. It was easier than what he had expected with an aircraft that size. As he climbed away, he looked back over his shoulder and got a set of hand directions on which direction to go. He turned into the direction and continued to climb. He had not really been told what altitude to go to and all he had was the hand direction from his instructor pilot.

They flew around for a bit. Brian watched his instructor, followed Esling's hand signals and after a half hour he received a hand signal that they should go back in and land. He returned to land and looked for the green light signal to land from the control tower but did not get it. He pushed the throttle forward to make another circuit around the airfield. Brian looked back at his instructor, who more frantically gave him the signal to land. Pilot Candidate Drummond figured that the procedures were different, lined himself up into the wind and came in to land.

As he landed, the instructor moved his aircraft ahead. Brian figured out quickly that he should follow the Esling back in, which he did. Taxiing in to the parking area, Brian used the same procedures that Malcolm had taught him learned at RAF Brize Norton. Brian parked the aircraft, closed the throttle, closed the mixture, turned the magnetos off, shut the battery power off, and got out of the aircraft. By the time he had descended to the ground, Esling was standing there in front of him.

"You bloody idiot, don't you know about radios? Why didn't you turn on your radio?"

Brian staggered. "Radio?" He had never used a radio. The Tiger Moths did not have radios, nor the aircraft he had flown in the U.S. with Malcolm. He had not even thought about a radio. In fact, the Gloster Gladiator was the first aircraft he had ever flown with a radio. He started to explain but knew that no explanation would satisfy an angry flight instructor.

"Get back up into the cockpit," Esling commanded.

Brian did as he was told and knew he was about to get a lecture about the radio. Flying Officer Esling described how to turn it on, how to tune it, and what crystals were, to make sure they were all on the same frequency. Brian listened intently to the radio procedures they would use.

"Do you think you could manage such a simple box?" he asked.

"Yes, sir."

"We are going to see if we can try this again, and do it the right way. This time make sure your radio is on after you move the aircraft and make sure you turn it off before you shut down the engine. Is that clear, Mister Drummond?

"Yes, sir."

"Well, then let us get to it, shall we?" Esling said angrily.

The procedures were repeated just as he had done before. This time he got the radio on and listened intently as his instructor came up on the radio and asked him if he was listening.

Brian keyed the switch as he was instructed and responded, "Yes, sir."

"Then, we will proceed."

Brian listened to the radio call as his instructor called the tower for clearance to taxi. The tower returned with the wind direction and speed, and the barometric pressure setting. Once again, he followed the instructor out to a taxi line. Esling asked him once again if he was listening. Brian indicated that he was. Esling called the tower for clearance to take off, which they received.

Flying Officer Esling radioed, "Mister Drummond, you are cleared for take off. You may proceed."

Brian came back with the appropriate, "Yes, sir."

He repeated the takeoff process just as he had done before, since there had been no complaint about his technique, he figured that he had done it reasonably well. So, he concentrated again to make sure everything was nice and smooth.

Once they were airborne and clear of the airfield, Esling broadcast, "Mr. Drummond, climb to two thousand feet, heading one five five degrees."

"Yes, sir."

Flying Officer George Esling directed Brian to perform a series of maneuvers. Gaining confidence in the aircraft and his ability to do it, Brian flew a set of turns increasing angle of bank, and trying to hold the aircraft at a level altitude. All the maneuvers he had practiced many times with Malcolm, Brian went through the whole sequence as described, with Esling following along behind his wing. After completing all the maneuvers and establishing straight and level flight, Brian waited for the next instruction.

"Mr. Drummond, we will trade places now. I want you to follow on my tail, do not get too close, do not get too far. I want you to do exactly what I do and follow me through each maneuver. Is that clear?"

"Yes, sir."

With that Brian maintained straight and level as Esling moved to a position in front of him. Brian felt a little bit close, backed off the throttle slightly to increase the distance, but then, remembered Jeremy Morrison's instruction. He moved back into position as Jeremy had taught him. Esling soon began a series of turns, climbs, and descents, increasing in rapidity and angle. Brian followed along quite easily as he had done many times with Malcolm and others in an exercise Malcolm had called 'tail chase.'

Watching their speed build, Brian recognized that Esling was probably setting up for some aerobatic maneuvers. Brian prepared himself, checked his instruments, checked his harness, and sure enough as the airspeed increased, Esling pulled straight up into a loop. Brian followed him directly, added throttle to stay with him, as they climbed up over the top, and began to get closer and closer as they went through the inverted position. He backed off on the throttle, held his position reasonably well, and as they came back down the backside of the loop, he had to pull the throttle back nearly to idle as they descended through the straight nose down position.

Once back straight and level, they held the position for a while, then Esling pulled up and began to roll in a barrel roll. The smooth attitude change made it much easier from Brian's perspective to maintain position as he followed him through the maneuver. They did a series of aerobatic maneuvers with Brian following. Trying to maintain position was not always easy as Brian tried to anticipate each move. He had never really gone through all of these different sequences so close to another aircraft before, but the techniques were just like Malcolm had told him, and as refined and polished by Flight Lieutenant Lord Jeremy Morrison, Esq.

The sequence of maneuvers that they had gone through took the better part of an hour and a half. At that point it looked like they were concluded.

"Mr. Drummond, return us home, if you will."

"Yes, sir."

Brian looked around at the features of the countryside passing below them. It did not look familiar. All of a sudden, he had realized that he had not paid much attention to where they were. He had been concentrating on Esling's aircraft trying as hard as he could to maintain position. He quickly looked at his compass. They had headed to the southeast over mountainous terrain. Brian deduced they must have been working to the southeast over the Welsh mountains. If he headed northwest, maybe he could get back to the general area. RAF Hawarden Aerodrome was at the end of the River Dee estuary just south of the River Mersey estuary with the prominent Birkenhead peninsula between them. He knew if he could get close enough to see the water, he could probably figure his way back to the airfield.

He headed to the northwest waiting and looking. The mountains gave way to gently rolling farmer's fields with the broken irregular lines that had become so familiar. He had been going for about ten minutes and was beginning to question whether he was heading in the right direction or not. He looked back over his shoulder. Esling was right there on his wing. No expression, no signs, no indication.

About the time he was ready to ask for some assistance on which way to go, Brian noticed the tell-tale signs of the Mersey estuary appearing through the haze. Quickly he oriented himself on what he had seen after takeoff and tried to locate the aerodrome. It took sometime, and he had to take a rather circuitous route, but he soon spotted the characteristic hangars and buildings of RAF Hawarden.

He waited for a call from or by Esling, but received none. He figured since he was leading it was probably his responsibility.

"Hawarden tower, this is Don Flight Charlie with two aircraft southeast for landing," he figured he would say as he had other pilots say in various conversations.

"Don Flight Charlie, wind is two eight five at seven. There is one aircraft in circuit turning crosswind. You are cleared to enter following that aircraft."

Brian spotted the aircraft in the landing pattern, positioned them well behind it as he had been taught, turned into the wind and landed. Without any other instructions, he taxied to the D Flight area and parked. Esling followed. Brian waited for Esling's engine to stop, and then he shut down his engine, making sure to turn the radio off first. After securing all the switches, he got out of the aircraft. All he received from Esling was a hand motion for him to follow back to D Flight Operations building.

George led him into a separate room – the same room that they had used to brief before the flight. Esling pulled out a card from his flight suit and set it on the table along with his flight kit.

"So, you think you are pretty good, do you now, Mister Drummond?"

Brian was not quite sure how to answer the question. "No, sir," he said trying not to be too confident or too arrogant in his response.

"Well, I must say, you exhibit fairly advanced skills for someone at this stage of his flying career, but you still have an awful lot to learn."

They talked about the flight. Once more he got a very severe lecture on not having read his Pilot's Notes or known that the radio was part of the procedures, stressing that his understanding and execution of the procedures were probably one of the most important things to save his life in combat. While that advice was certainly not the same advice provided by Malcolm Bainbridge, or Jeremy Morrison for that matter, he knew that Esling meant what he said. He also knew that he had to pay particular attention to procedures while he was at advanced training.

They went through the sequence of events – the maneuvers – they had flown during the second sequence of the flight with the radio on. Esling described some of the subtleties, some of the techniques that Brian had used and where he might do better. George was clearly impressed with his flying skills, but wondered whether he would have the ability to perform within the rules that had been established for the Royal Air Force and more specifically for Fighter Command.

Esling emphasized teamwork and the fact that the team was more important than the individual and that the team depended on each member doing exactly as he was expected to do, exactly as he was told, so there was no question, no confusion on what to expect. The words that were used in the description were very much contrary to everything he had learned before.

He had learned individual skills and achievement in hunting techniques, methods to stay alive. Malcolm had told him nothing about team flying and flying wing for a leader. He was not sure how to take it. Was Malcolm right and these procedures wrong? Should he do what he had been taught, or do as he was being instructed?

The decision came quite easily, although the conflict within him was significant. For at least the time he was at RAF Hawarden, the procedures were apparently the key. He had to concentrate, whether he agreed or not.

After further discussion on the ground about the specific tactics in operations that they would be learning in the near future, the process of navigation, the radio procedures, the techniques for dealing with foul weather flying and of course the tactics of fighters. They talked about guns. How to use them? How to care for them? How to know when something is right or wrong? It was all filled with so much information that afternoon; it was almost too much for Brian to handle.

It was some comfort though when he had a chance after the evening meal to compare notes with Jonathan Kensington, and what he had gone through. Although very much the same, Brian was quite conscious of the fact he had done more, in more severe maneuvers, than what Jonathan had done. He kept it to himself. Brian certainly did not want Jonathan to feel he was better, had done more, had learned more, or had been exposed to more. Brian was very quick to draw attention to his biggest mistake, which was not knowing about the radio. All the student pilots of D Flight got a big laugh out of it. They enjoyed it, and they called it an early night.

For the remainder of the evening before he went to bed, Brian studied the Gloster Gladiator Pilot's Notes, read them through four and five times very carefully learning each sequence. As he committed the procedures to memory, he recognized he had handled the Gladiator correctly for the most part, but there were several things that he had done wrong and had not been spotted by his instructor. He vowed not to make the same mistake again.

―

Saturday, 15.July.1939
Liverpool, Merseyside, England

THE first week of advanced flight training for both Brian and Jonathan proved to be much more difficult than either had imagined. The expectations, as well as the demands that were greater on both of them, made it seem as if the first week lasted more than a month, yet the calendar showed five days. As they learned on Friday, the weekends were free time although rumors persisted about training seven days a week. The war jitters affected everyone, but for the moment, Operational Training Unit No.7 continued through a normal routine. As Brian and Jonathan learned from their brethren, the lights and night life of Liverpool had become a tradition for candidates of OTU7 seeking female companionship, entertainment, pleasure, and sometimes trouble. Another element of tradition was that the candidates of each particular flight usually went in or at least started the weekend together.

Although Brian was not particularly enthusiastic about a weekend in Liverpool, he knew that camaraderie with the other pilots was also important. Left to his own considerations, Brian would have remained behind and studied a number of the tasks and techniques he was having difficulty with. The pressure of being part of the group, being part of the team, overcame his personal concerns.

The other candidates Brian met over the course of the week came from a wide variety of backgrounds. None were members of the aristocracy like Flight Lieutenant Lord Jeremy Morrison, Esq.

The others ranged from James Roland, the oldest son of an ironworker in Coventry, to David MacGregor, a graduate engineer from Glasgow. Derek Langston had grown up on his father's fishing boat until the years of watching various aircraft flying around Southampton and the impending war brought him into aviation. Ian Milner was a quiet, shy and reserved young man not much older than Brian who grew up on his parents small farm in East Anglia.

Brian was the youngest and only non-British citizen in the flight. As such, he received different forms of support, as well as tailored jokes and ridicule. He sensed certain resentment toward him from a few of the candidates. He could never really determine whether it was because he was an American, or because his reputation as a good pilot put him on a different level than the others.

The natural gravitation among all the candidates of D Flight was towards Jonathan Kensington. He seemed to be the friendliest, the most concerned, the most open of all the other candidates. It was easy to like Jonathan. Fortunately for Brian, the friendship between them grew by the day. Information, advice and counsel came freely from Jonathan who helped Brian repel some of the resentment directed at him.

As the two new members of D Flight, they learned the internal expectations within the group, among which was the recently proclaimed tradition that all candidate pilots of D Flight travel to Liverpool as a group for their Saturday entertainment and pleasure. This particular weekend represented, as Brian learned, more the norm than the exception. Four of their number remained behind due to various forms of restriction, all except one needing extra instruction with their flying lessons. Their usual guide through the night spots of Liverpool, Peter Baker, was this day one of the restrictees.

"What is it going to be tonight, lads?" asked Pilot Candidate Ian Milner.

"I'd say we should have a go at the Mermaid and Serpent," James Roland answered.

"Yes, indeed, that should do."

The small band of fledgling RAF fighter pilots followed James Roland through the streets of Liverpool from Lime Street rail station toward the Strand. The streets were narrow, loud and crowded with traffic and boisterous patrons of corner pubs. The evening was quite young with the summer sun still well above the western horizon although the dark shadows of the close quarters made it seem later than it was. Brian continued to adjust to the extended hours of daylight in the higher latitudes of Great Britain. Several intoxicated burly men staggered up the street having reaped the benefits of alcoholic bliss.

"Watch yourself," Jonathan whispered to Brian. "I suspect we are venturing into a rough section of the city."

Brian nodded his acknowledgment without truly knowing what he would do having never really been involved or even close to the violence often common to drunken men.

The docks along the River Mersey filled the vista through the surrounding buildings as the large sign over the door proclaimed their arrival. The buxom, bare chested, blond haired, mermaid holding the head of a large snake wrapped around her midriff on the sign did not need any words to prove the name of the establishment. Good times were well underway from the sounds of laughter and indiscernible shouts from the interior. They all followed James into the pub. The stale odor of smoke and beer permeated the interior as it did so many similar establishments.

"Our very own junior birdmen are back, lads," came the resounding announcement as they entered. "Ready for another go at the beer kegs, are ya?"

"Without question," answered Pilot Candidate James Roland leading the small RAF contingent to the bar. "Beers for me mates, if you please." The chorus of cheers confirmed the satisfaction of the patrons with the action.

"Where's our own wee Peter Baker?" asked the boisterous man.

"Afraid he's been detained," responded David MacGregor.

"Oh well, I'm sure he'll catch up. Always does."

A couple moved to another table to make a place for the six pilot candidates. The large round table gave them all plenty of room from the corner to the side of the bar in the main room. It was difficult to sort out who was with whom, but each of the young men recognized the reality of a nearly equivalent split in gender. Hunting would not be good, several of them acknowledged. It took less than one pint for James to identify his first target.

"Excuse me, gentlemen. I have work to do."

"My God, Jimmy boy, can't you at least give the lasses a couple of pints breathing space?" asked MacGregor sarcastically.

"Time passed is time lost as far as the dolly-birds are concerned."

"Some of these may not be dolly-birds, Jimmy."

"We will find out soon enough, then won't we," he said as he stood, adjusted his uniform tunic and walked away toward a small cluster of young women.

"Watch this," Derek Langston suggested looking directly at Jonathan and Brian. "If any of those delicate chickies are unattached, or otherwise available, he will be out of here within ten minutes."

Brian looked to Jonathan who simply shook his head in disbelief. The talk quickly jumped to flying. Both the new joiners kept an eye on James as they answered questions about the trauma of the first week of advanced training. It had become another bit of ritual since the Gloster Gladiator showed up in the Training Command. The rumblings of war added to the pressure, tension, fervor and seriousness of advanced training. The next step after completion and being awarded the distinctive crowned wings of an RAF pilot was an operational fighter squadron for most of them. Brian began to take note of the professed desires of the pilot candidates around him. There was a number, larger than he expected, who had absolutely no yearning for the confinement and isolation of a single seat, fighter cockpit. The contrast for Brian was quite clear. The singularity of a fighter in the sky was probably what had attracted him most listening to Malcolm Bainbridge's stories of aerial combat during the Great War.

"Wait," said Derek holding up his right hand looking beyond the group.

They all joined his line of sight.

"There he goes. What did I tell you," he added with a voice of satisfaction.

James Roland was indeed walking out of the pub with a rather large but not overweight brunette on his arm. The woman was nearly a head taller than James. Neither seemed to be particularly concerned about the difference as they disappeared into the dusk.

"I suppose we won't see him again," observed Jonathan.

"Oh, quite the contrary, my good fellow," answered Derek. "Depending on the sequence of events, our good friend could be back for seconds within an hour or so."

"Then again, maybe not," added David.

Several Royal Navy sailors joined the jovial group. An obvious familiarity existed between many of the patrons. Less known patrons and even strangers were readily accepted. There was talk of war, but from what Brian could pick up, the words were mostly of the sea and recent experiences at the hands of Mother Nature. Stern, serious expression and hateful words accompanied the brief, angry discussion about U-boats, and what war meant to the sailors. Brian remembered the cautious advice about Liverpool being a seafaring town. When his thoughts were allowed to drift, images of Anne Booth mixed uncomfortably with the still clear memory of Rebecca Seward. He missed them both for different reasons. The freshness of Anne commanded the dominant capacity of his thoughts. The clarity of his images proved to be sufficient for Brian to dissuade him from joining the game the others began to play.

In the span of thirty minutes, only Brian and Jonathan remained at the table. The others circulated throughout the pub, each looking for the pleasure of female company with Ian Milner the most reluctant, but not particularly resistive participant.

"This is not my cup of tea," Jonathan offered.

"Nor mine."

"You want to go before any excitement starts?"

"What do you mean?"

"Mixing seamen and airmen has never been a particularly successful endeavor. Furthermore, with some of the expressions on the locals, I'd say we may see a bit of a row before long. The women appear to be quite content being the center of attention, while more than a couple of the men are not especially impressed."

Jonathan's observation jumped into instant focus once articulated. A sense of growing irritation was clearly perceptible. Jonathan's warning on the walk down from the rail station came back to him. Maybe it was time to leave.

"I'm with you," announced Brian.

"Good, then, let's go before something nasty starts."

As they stood to leave, a burly, hairy, tattooed man reached his limit. "You bloody fly-boys have gone too far. You leave me damn woman be," he shouted in a heavy, threatening voice.

"The lady has not yet indicated she is attached, my dear fellow," said

David MacGregor.

"Well, she is, and I have a mind ta knock yar bloody 'ead off."

"No need to get hostile."

"Hostile, I'll show yous hostile."

Before the confrontation could come to full bloom, Ian Milner jumped in. "Come on, David. Let's go."

"Why?" he answered pulling his arm from his grasp. "I think this young lady would rather be with me than this foul smelling bloke."

The derogatory comment pushed his antagonist over the edge. He stepped forward and took a swing at David. A precise, sufficient bob of his head avoided the blow.

"Let's go, now, lads," commanded Ian.

They all moved to the door including David who was the most reluctant. The burly man with clenched fists followed them as if to ensure they did not stop. Out on the darkened street, the mood changed.

"Why did you do that?" asked David, not very pleased. "I could have handled that oaf."

"Perhaps. Then, the rest of us would have been enjoined when his colleagues came to his rescue."

"We could have dealt with them."

"Tell me, David," said Ian, as they walked toward the station, "why is it you feel this compelling need to fight over women?"

"The enjoyment of female flesh."

"Is that all they are, a toy to be enjoyed?"

"Of course. We are all going to die in the coming fracas anyway. Why not take pleasure from the fruits of life while we can."

The somber comment iced any additional conversation for the remainder of the evening. At the station, David and Derek decided, in the end, to stay in Liverpool for the singular purpose of finding some bedroom companionship. The remainder of the diminished group made their way back to the aerodrome without further discussion.

―

Chapter 8

> It is better to die on your feet
> than to live on your knees.
>
> -- Dolores Ibarruri

Saturday, 15.July.1939
Liverpool, Merseyside, England

"WHAT a day," exclaimed Brian as he walked back into the D Flight residence building.

"That bad, was it?" asked Jonathan Kensington.

"Esling just wouldn't get off me. He kept after me all day long. I'm still having trouble with this damn formation task. He keeps giving me these strange conditions, and I'm still not familiar with the alignment cues on the Gladiator. I keep backing off which causes a cascade of other problems."

"Welcome to the world of fighters," offered Derek Langston.

"Is Flying Officer Esling always so hard to get along with?" asked Brian.

"He certainly is. He has not had an operational tour, as yet, and I think he enjoys taking out his frustrations on all the rest of us," answered James Roland.

"Well, I've got to figure out how to satisfy him and get him off my back."

"I do not think you ever will. He enjoys making life difficult for new pilots, especially young, cock-sure, pilots with inflated reputations," added David MacGregor.

Brian's frustration was brought nearly to the point of reaction, but there was no reason in reacting to the pointed taunt. "Well, I've still got to find a way," Brian mumbled.

The warmth of the summer day compounded Brian's level of fatigue. He had worked hard and was still damp from the perspiration generated during the flight. The windows were all open, but there was barely a waft of breeze to move the air. The smoke from David's and Peter's cigarettes and Derek's pipe did not help his mood in the slightest. He wanted to take a cold shower to cool off, and then relax a little before the evening meal and study period.

"Well good, then maybe you can take your mind off all your difficulties for the day. The quartermaster delivered the mail. It seems that you're a bit popular there, Mr. Drummond. You have three letters here waiting for you."

It was the first mail, other than the telegram from Malcolm, Brian had received since leaving home nearly six weeks ago. The mail had finally caught up to him. The three letters were from Malcolm, his parents and Becky.

Brian stored his flight kit, took his tie off and sat down on his bed. He stared at the three letters on his lap as if he were in a debate about which one to open, but there was no debate. He would open Becky's letter first, but he had difficulty making the first move. He wanted to see what she had to say, what thoughts were on the paper. It was the difficulty of facing the possibility of rejection as well as a clear reminder he missed her a great deal.

"Are you just going to sit and stare at those letters forever?" asked David in a very sarcastic tone, "or do you need someone to read them for you?"

Brian looked up and scowled at David. His gaze returned to the three letters. He picked up Becky's letter and opened it carefully.

June 10, 1939

Dear Brian,

I hope this gets to you somehow. I got the address from your Mom. I still can't see Mr. Bainbridge because I blame him for taking you from me. Anyway, I hope you get this. I've never written a letter to someone outside Kansas.

I have so many questions. I don't know where to start. I love you, Brian. It has only been a week since you left, and I miss you so much I just want to die. Why did you do this? Why did you leave without saying anything to me, without a kiss, without even a good-bye? Your stupid little note just made me more angry, more hurt, more upset. Is flying that important you want to kill yourself?

I gave myself to you because I love you and I thought you loved me, but now I see you love flying more than me. It hurts, Brian, it hurts a lot. I wanted us to be together, to go to KU together, to get married, to have children. I wanted so much for

> us, but now I have nothing. You have gone off to England and this damned RAF thing. I don't even know what RAF is!! Sometimes I think you wanted to get yourself killed for some reason. I know you like flying, but I don't understand why you like flying more than me, more than what we have together. Didn't you enjoy our making love? Don't you miss it? Don't you want more? I sure do, but I can't without you. You are the only person I have ever even dreamed of being with.
>
> I've got so many more questions, but I don't even know if this will get to you. Mrs. Bainbridge says her husband sent you a telegram, but they haven't heard from you. A friend of Mr. Bainbridge sent a telegram back saying you are safe and on your way to England and you should arrive in a few days. You should already be there if you get this letter. Please write to me and tell me what is happening and why you are doing this to me. Enough for now, I'm crying too hard to see the words anymore.
>
> Don't you ever forget, I love you. I want you home as soon as possible!
> Love bunches,
> Becky

Reading the letter made Brian feel small, selfish and depressed. He still regretted his departure method because he did care for her so much, but he also knew that he really had no choice. The difficulty of being away from home, away from friends and family, came home quite harshly. There was relief in the words Becky used. She still loved him, wanted him to be home, and missed him a great deal. The sentiments were very much the same for Brian although from a distinctly different perspective and now there was a new dimension for him. In some ways, he wished he could undo that Saturday night a week and a half ago. In many more ways, he knew he could not or did he want to. Brian wondered how his experience with Anne would change his feelings for Becky. Time would tell.

He placed her letter back into its envelope and opened his parents' letter.

June 8, 1939

Dear Son,

Mr. Bainbridge tells us this letter will get to you so we will have to trust that it does. To say the least, we are disappointed, angry and hurt by your actions. Leaving us with just a feeble note was not fair, Brian. It makes us regret so many things, but most of all letting you associate with that damned Mr. Bainbridge. He poisoned your mind. We hold him responsible for what you have done although the law may not be able to. What he did and what you did were dreadfully wrong, Brian. You need to come home right away.

We have talked to the police. They can't do much. Apparently, Mr. Bainbridge did not violate any laws in helping you. But, we have talked to the FBI and you have violated several federal laws, Brian, specifically the U.S. Neutrality Act which prohibits any American citizen from joining or fighting for the armed forces of a foreign nation. They are looking into what they can do to return you to us. We have also written to Congressman Houston and both Senators Capper and Reed for their help in ending this foolishness and bringing you back to us.

What you are doing is wrong, Brian. If there is going to be a war in Europe, it is their war, not ours. They don't need the help of a young boy from Kansas in fighting their wars for them. I know you must think that what Mr. Bainbridge did in the

> Great War was heroic, glorious and something to be admired, but it was not. He was just as foolish as you are being now and you may not be as lucky to survive it as he was. Please, Brian, you are our only son, our only child. You simply cannot do this to us. You must give up this stupid quest of yours and come home immediately. We don't want you to die in some foreign land for some cause we have no part of. You belong home with your family. Becky misses you very much and was also very hurt by what you did. All will be forgiven if you come home now.
>
> We miss you very much, Brian, and we fear for your safety. Please come home, now.
> Love,
> Mom & Dad

In several ways, Brian felt defeated by his own parents. He sat on his bed like a pile of mush, his strength drained from him. He could only shake his head with his own frustration over people not wanting him to fulfill his dream. Why did some of the people he cared most about not understand his burning desire to fly the best aircraft and contribute to winning the coming righteous struggle? He could only tell himself ... parents were parents. They simply feared for his life. They wanted him safe. For that he was thankful. However, he would certainly feel much better with their support. The most important questions remained, would they actually be able to force him to give up his dream and return to America? Brian recognized more than ever he should have written to his parents when John Spencer asked him upon receiving Malcolm's telegram. He had to write now, or he might be denied his dream. Maybe Malcolm had some good news for him.

```
June 11, 1939
Brian,
      Please forgive the formality of a
typewritten letter, but I asked Mrs. Bainbridge
to type this letter, so you would have no
trouble reading what I need to say.
```

First, I suspect you have received, or soon will, letters from your parents and Becky and maybe the government. Your joining the RAF has caused quite a stir. It has been in the newspapers for days. The general opinion goes along the line that I corrupted your young mind getting you involved in a foreign affair and risking your life if there is a war in Europe. Most people don't understand just as they did not when I joined the RFC in the Great War. The only thing I can tell you is, you must search your mind and your heart. Are you doing what you really and truly think is right and you want to do? The most important thing is what is in your heart. If you are not absolutely convinced, you will not have the edge and conviction you will need to endure the ordeal. You must be sure.

Second, I believe your parents intend to initiate some official action under the Neutrality Act to force you to return to the US. I have no idea what you can do about this. It is real, Brian. The mood in this country is very much stay-out-of-this-one. The only advice I might offer is to talk to John Spencer. He is a good man and should be able to help.

Third, I want you to know I believe in what you are doing. There are others here who also believe in your noble effort. I just wish I was younger and I would be with you. What Hitler is doing in Europe is dreadfully wrong even if most of the world cannot see it as such. If you are convinced you are doing the right thing, then you are!! Have faith.

Lastly, I know you do not need any of these burdens now or in the future, but as we discussed this Neutrality Act is real. Do not take it lightly. Above all else, if you remain in the RAF, you must devote

> all your concentration and effort to your flying. Distractions are not good and must be eliminated from your thoughts. As I have told you many times, air combat is nasty business. Things happen very quickly and even a moment's complacency can mean death. You must concentrate, Brian.
>
> Good luck. Keep your eyes moving. Check six! You'll be OK. Say hello to John. Trust him, he knows what he is doing. Write when you get a chance.
>
> *Malcolm*

RAF Pilot Candidate Brian Drummond sat on his bed like the pile of dung he felt like. The three letters plus Malcolm's original telegram left him so alone with nothing but obstacles, chasms and raging torrents all around him. The waves of defeat crashed about him as he wondered what he should do. How could he fight the government and even more importantly, how could he overcome a demanding instructor while faced with all this resistance and people pulling him down? Malcolm seemed to be his only source of strength. He knew John Spencer would help, but he was a very busy man.

While Brian clearly recognized the thoughts since he was in the middle of it now, it was intriguing to read Malcolm's perspective. It also brought a smile and a chuckle as Malcolm reiterated some of the important lessons of aerial combat and survival in the air. As he read the words, he wondered if he would ever get an opportunity to actually practice the lessons he had learned.

Brian was glad he had saved Malcolm's letter for last. It was the most upbeat. The first two were quite sad. He had brought so much disappointment, discomfort and regret to Becky and his parents. He could sense the pride in Malcolm's letter, however, and for that he was thankful. In his own way, he was convinced that Becky and his parents would eventually come around to understanding and accepting his commitment.

"Well, what is the news from America?" asked David MacGregor.

Brian stood to take his shirt off and wondered what he would say to answer. "Just from my parents and my best girl, and from my flight instructor who's still giving me lessons on what to do."

"Isn't that sweet," said David.

There wasn't much movement in the heat of their barracks. What movement there was came sluggishly. The thoughts about the future compounded the heat-induced lethargy.

"Well, it's good to know that some things are near normal," added Ian Milner.

"It certainly is," said Derek. "My uncle is a sergeant with the BEF in France. They have been put on an increased state of readiness. Although he can't give us much detail, it sure does seem like he believes something is going to happen soon."

"We are seeing all the indications," said Peter Baker. "Just read the newspapers and listen to the BBC. It's like they're trying to say things are OK and perfectly normal, yet everyone keeps coming back to what the Germans are doing."

"I don't think anyone really knows what the Germans are doing," observed Jonathan.

"Well maybe not, but my uncle is concerned about things. He's asked my aunt and my father to prepare themselves in case war should occur."

"Everyone is talking about Hitler, the Nazis and Germany. No one seems to know quite what to expect from them," said Ian.

"Tell me young Mister Drummond, are you going to run home to mamma when the shooting starts?" asked David MacGregor.

"No more than you will," answered Brian. "I'm here to do the best I can."

"But, this is not your fight."

"Maybe not, but it is still a fight for freedom, and I intend to add my skills to the side of good."

"My, my, lads, we may have a wee bit of a righteous philosopher in our midst," said David MacGregor.

Was David just being argumentative, or was he really so pessimistic? David was the darkest of all the D Flight candidate pilots. He always seemed to see the darker side.

The topic of war and what was happening in Europe never really seemed to get away from them very much, unless they were around women. Everyone seemed to have an opinion, although Brian could never find very much real evidence that anything was happening. The events in Europe were of course an important interest to all of them since they were training in a profession that would put them in the forefront of any combat action. With only a few exceptions, each of them had their minds on the task at hand, to pass the next flight check in order to get their wings and join an operational squadron. It was the flying that attracted each of them in their own way.

Brian had some visions of right and wrong, but it was clearly the flying that was most important. While the Gloster Gladiator was not the fastest airplane in the sky, it was faster than most and certainly more maneuverable. It was faster than Malcolm's Stearman, although not quite as fast as the modified Mystery S he had flown in St. Louis. It was still an impressive aircraft to handle. You could feel the power. And it was a fairly nimble machine, more so than the Grumman FF-1 he flew with Group Captain Spencer in St. Louis. The discussions often, in fact invariably, turned to flying, if it was not the approaching war. Each of them enjoyed flying in their own way, some for many different reasons. It was always a good form of reinforcement to be around other men who enjoyed the same interests.

The others seemed to break off into separate discussions about politics and the way the government handled activities with respect to the European situation. There was also a form of arrogance that history protected Great Britain because of its existence as an island. While wars may be fought in France, or Germany, or Austria, or Italy, the United Kingdom had never been seriously threatened since the weather and the Royal Navy thwarted the Spanish Armada in 1588.

The build-up of the armed forces, talk of war and the seriousness of the situation in Europe, always seemed to show up in conversations, but was held at arm's length since most Britons did not seriously consider any risk to Great Britain. More often than not the voice and words of the Winston Churchill would come up, sometimes jokingly or in a derogatory way. So they would say, only Churchill believed that war could touch Great Britain. Occasionally, some conversations, whether military or not, referred to him somewhat like a prophet. They believed he was correct. He was forecasting events to come and the need to prepare. More often than not, however, the speaker would refer to him in a rather defamatory way, calling him a saber-rattler, a war-monger, a doomsayer and many other terms indicating the speaker's general disbelief with the message that Churchill had brought to Parliament and to the British people. Brian remembered the stories and words Malcolm Bainbridge had told him, increased and enhanced by the personal knowledge of John Spencer.

Brian tried to question what was right and wrong, but he knew there was only one thing he truly had to worry about, doing the best he could at flying. Malcolm's words constantly reminded him. It was also good to know his new friend, Jonathan Kensington, believed in the words and the predictions of Churchill. It was easier to talk to Jonathan about the situation and the issues. Listening to David MacGregor, you would think the world was about to come to an end. Somehow all the other nonsense about the Neutrality Act, repatriation and the forced denial of his dream did not seem to matter. Those additional obstacles would be overcome as had those previously put before him.

Friday, 28.July.1939
RAF Hawarden
Broughton, Clwyd, Wales

A sea of RAF blue uniforms flooded the officer's mess bar. Several wives or girlfriends added a splash of color to the weekly graduation celebration. Seven young men had just received their pilot's wings. As was usually the case, they retired to the bar to celebrate and congratulate their colleagues for the accomplishment and to send them on their way to operational squadrons.

"I didn't realize there were this many pilots at Hawarden," remarked Brian, speaking only to Jonathan Kensington.

"Likewise. I have not even seen most of them before."

"Well, maybe we can have a few beers, shake David's hand and be on our way. What do you want to do tonight?"

"Actually, I'd just as soon call it a night, do some studying and get to bed early," said Brian.

"Then, I may try to go home this weekend, rather than stay around here."

"That would be great."

"Would you like to come with me, Brian?"

"Of course I would, but I've got to study several skill lessons. I start gunnery, and if I'm not ready for Monday morning, Esling will eat me alive."

"Oh, come now. You can't keep pressing on like that. You need a break. Come up to my family's home, and you can relax."

"No, thanks, Jonathan. I appreciate it. Maybe some other time. I really do need to study. There's only one reason I'm here and that's to fly. If I can't get that right and get the assignment I want, then I have no purpose for being here."

"You take things far too seriously, Brian. You must loosen the collar now and then."

"I am, I'm having a bitter. That's enough isn't it?"

"Well, maybe so," said Jonathan without resolve. "Let's go say good-bye to David, congratulate him and maybe we can be gone."

"That sounds good."

The two young pilots worked their way through the crowd, listening to the various conversations of their colleagues from other flights. The words were all related to aviation and flying.

Upon arriving, David was talking with several other officers, a couple of instructors along with Peter Baker. As Brian and Jonathan joined the small group, they realized the topic of discussion was David's posting. He proudly stood with his shoulders back as if to emphasize the new wings of an RAF pilot over his left breast pocket. Brian and Jonathan listened as the veteran officers discussed David's new posting. He was assigned to No.263 Squadron which was being formed at RAF Filton near Bristol in the Southwest and equipped with new Gladiator fighters. While David seemed particularly excited to be part of a newly forming unit and certainly pleased with flying an aircraft he knew, it was difficult for Brian to maintain his attention on the conversation before him.

He heard another conversation, another discussion going on behind him, which gathered his interest to a far greater extent. As he listened it was a similar discussion going on apparently between members of A Flight and several of their instructors. One of their colleagues was being posted to No.19 Squadron at RAF Duxford near Cambridge flying Spitfires. The newly designated pilot was obviously excited and enthusiastic about the aircraft they flew and the prospects for their ability to deal with air combat with the new machine.

Brian would have much rather joined the conversation going on behind him, but he did not know any of officers and did not feel comfortable jumping into a conversation to which he was not invited. He accepted just the ability to hear snippets of the conversation and let his imagination consider the benefits of the fast, sleek looking fighter aircraft.

He leaned over and whispered to Jonathan, "Do you hear these guys talking about Spitfires?"

"Of course, I hear them."

"Wouldn't it be great to fly them?"

"I suppose."

"I sure want to. They are much faster than the Gladiators we fly."

"I suppose you are correct. It would be better as they have told us repeatedly, speed is life, so I suppose that's the best thing."

"I certainly don't want to get posted to a Gladiator squadron like David. Somehow I've got to figure out how to get reassigned to Spitfires either here, or when I get posted to a squadron."

"Well, good luck, Brian. It's not clear to me how often pilots are able to make this switch. I suppose you would have to know somebody in order to get assigned to Spitfires."

"Maybe so, but I'll figure out a way."

"I know you will, you are fairly determined," said Jonathan who was obviously loosing his interest in their conversation. "Why don't we say our congratulations and leave. I should be on my way home anyway."

"Sounds fine to me."

Jonathan took the lead and waited for the appropriate moment in the conversation. "Congratulations David, we hope to join you soon. Good luck at your new posting and enjoy your accomplishment."

"Thank you, Jonathan," said David with his thickest Scottish brogue.

"Congratulations as well," said Brian.

"Thank you young Mister Drummond, I appreciate it and good luck to you. I hope you figure out how to fly formation in the not too distant future."

Brian did not appreciate the offhanded comment about his difficulties with formation flying, but as he had done so many times before he chose to let it pass. "I'll figure it out in time," he answered with as much politeness as he could muster.

"I am certain you will. Thank you and good luck," said David.

With that, the two candidates began making their way towards the exit. As they weaved through the clumps of fellow officers and a few ladies in the officer's mess bar, they made it outside. Several small groups of pilots were talking, holding several pints of beer at various levels of consumption. The topic of discussion was unmistakable.

As they walked Brian thought about what he had just heard and he asked Jonathan, "Why aren't you excited to fly Spitfires?"

"I don't know, it is just another aircraft."

"Just another aircraft," exclaimed Brian. "It's not just another aircraft. I mean all you have to do is look at it, look at the lines. It's probably the most beautiful aircraft ever built. It certainly is the fastest, or the fastest I'm aware of. Well, maybe other than the specially modified racer aircraft. But, it's certainly the fastest operational aircraft. I just don't understand why you're not excited about trying to fly it."

"Well, I suppose I am, but I just enjoy flying. Flying is flying. I will fly anything and enjoy it. Plus, I can fly and my father can't." A broad, sardonic smile floated across his face. "If I get assigned to Spitfires, I will enjoy Spitfires, but I have no problem flying Hurricanes or Gladiators for that matter."

"Well, OK, but it just looks so much better, so much faster, such a better looking machine and it simply looks fun to fly. The guys who fly them can't say enough good words about it. I mean how many people get excited about flying Gladiators?"

"I do."

"Well OK, I enjoy flying Gladiators too, but it's not excitement."

"Well, you take what you can get."

With that they continued walking in silence back to their residence area. The room was empty with all the other pilots down at the bar. Brian went to his desk, sat down, and made a move to get his study books out in preparation for doing a little reading.

Jonathan tried to complete packing up a weekend kit bag for his trip home. Deciding to take a break, he went into the orderly room and came back holding a letter. "Who do you know at fighter command?" asked Jonathan.

"Fighter command?" Brian thought for a moment and realized that it had to be John Spencer. "The only person I know at Fighter Command is an officer by the name of Spencer, Group Captain Spencer."

"Well, it doesn't say who it's from, it just says where it's from. Here you go, open it up let's see what it is. Maybe your getting an early posting."

The official envelope contained a simple slip of folded paper with block letters at the top, GROUP CAPTAIN JOHN SPENCER, DFC. Brian read the note quickly. "Well, it is from Group Captain Spencer. He's invited me to London next weekend." Brian thought for a moment wondering what the invitation might be for. "Do you want to go?"

"Brian, he is inviting you. He is not inviting me."

"He and his wife are real nice people. They wouldn't mind."

"Maybe for you colonials that's acceptable, but it certainly is not proper to show up without an invitation."

"Well then, maybe I should telephone him and ask for an invitation."

"Brian, Brian, Brian, that is not proper either so just let it rest. I am certain we will go to London sometime together, but you are not going to my home when I have invited you, so I think it is quite appropriate that only you go, since you are the one with the invitation."

"Very well then, I guess that's what I'll do."

"Speaking of that Brian, a group captain is a fairly high ranking officer, maybe that's your ticket to get posted to a Spitfire squadron."

Although the thought had occurred to Brian, he was still not quite comfortable asking John Spencer for that kind of assistance. He wanted to get assigned to Spitfires on his own, by his own accomplishment.

"I suppose, but I can't ask him just yet."

"Why not?"

"Well, I'd like to get posted to Spitfires on my own, because I've got the skills and the capability to fly them, and they think I'd best serve the RAF in a Spitfire squadron. But, I guess if I'd have to I could ask him."

"When you want something bad enough, you must do whatever you can to achieve it."

"Well, I suppose I'll have to consider it."

As Jonathan Kensington finished packing, Brian returned to wondering what John Spencer was inviting him to London for. It could be for a break from training. It could be something else. The note didn't say what it was about. The note simply suggested he should take the early London Midland and Scottish Railway train Saturday morning down to London. A room was already reserved for him at the Savoy Hotel. The image of Mary Spencer came to him. Maybe he would be able to see her beauty, again. Lastly, the note asked him to call the Spencer residence as soon as he arrived. It seemed like a reasonable plan. Brian put the note carefully into the corner of his desk drawer.

"Well, that's about it for me," said Jonathan.

"Are you ready to leave?"

"Yes, I think I will head on home straight-away. It will take me a couple of hours by train so I will miss supper, but I am sure I can get something to eat once I get home."

"Enjoy your weekend, Jonathan. I'm going to sit here and study the best I can."

"Well, maybe you should, but if you are going to do that, make sure you don't get sucked into going into Liverpool."

"Naw, I don't think I'm going to go this weekend."

"The others might be a little cross with you."

"Well maybe they will, but I'm here to fly, not go to pubs."

"Maybe so Brian, but flying is also camaraderie and teamwork. Never lose sight of that either."

"I won't, but I've got to get these basic studies down, or I'll never get to be part of the team."

"Well then work hard, and I'll see you Sunday evening when I get back."

"Enjoy."

Chapter 9

> If we had no faults of our own,
> we would not take so much pleasure
> in noticing those of others.
>
> -- Duc de la Rochefoucauld

Saturday, 5.August.1939
The Admiralty
Whitehall, London, England

"Very well, gentlemen. Thank you for your excellent efforts and sacrificing your time this glorious Saturday morning," Admiral Pike told his naval intelligence operations staff as much to congratulate them as dismiss them. He pretended to return to the appreciable paperwork on his desk as the four men gathered up their maps, charts and briefing papers. The product of the briefing had been clearly understood less than half way through the one-hour discussion. Several telephone calls in specific order had been agreed to a month earlier.

With the door closed, the large office quiet and enough time for the men to clear the outer office, Sir Geoffrey lifted the green telephone that had no dial and only one connection although two electronic devices scrambled and unscrambled the signal. After four rings, the handset was picked up.

"Yes," came the simple greeting as a matter of habit rather than disclosure who or what you were in some more formal or proper salutation.

"Good morning to you, 'C.'"

"And a delightful good morning to you, 'Jumper.' My apologies for not picking up promptly. I have Stewart with me."

Colonel Stewart Graham Menzies, DSO, MC, Deputy Director General of the Secret Intelligence Service – Sir Hugh's number two – was an accomplished and respected, career, intelligence officer seconded from the Army.

"I thought I might update you on our naval status, however I have more disturbing news we should only discuss in person and in private. Would you be so kind as to join me?"

"Geoff, I would love to, but I am afraid I am not so well. Might I impose upon you to come here?"

"My apologies, Hugh. I hope you feel better soon. I will leave immediately."

"Thank you. I would also ask to have Stew present as well. I have asked him to assume more of my daily activities."

"Surely, Hugh. He should here this as well."

"Then, we shall see you momentarily."

They hung up. Admiral Pike called for his driver, actually his armed bodyguard masquerading as an official naval service driver. The distance to SIS HQ was a modest summer walk across the corner of St. James Park from the Admiralty. The extraordinary times called for extraordinary caution.

—

Saturday, 5.August.1939
Headquarters, Secret Intelligence Service
No.21 Queen Anne's Gate
Westminster, London, England

ADMIRAL Pike entered the non-descript, former row house, now pressed into government service, alone. He passed through the requisite security procedures smoothly and quietly. He was ushered into Admiral Sinclair's inner-office. 'C' and his deputy stood. Gentlemanly greetings were exchanged before Sir Hugh motioned for them to sit.

"How are things at sea?" asked 'C.'

"The *Graf Spee* is headed toward the South Atlantic ostensibly on a goodwill tour of South America."

"Goodwill, indeed," interjected Menzies.

"Yes. We are convinced she has orders to interdict our trade routes as soon as hostilities commence."

". . . as soon as . . . ?" Stewart asked.

"Yes. The preponderance of signs point toward Poland in the next weeks or months at the outside, I'm afraid," responded Pike. "And, of course, our treaty obligations will require ourselves and the French to confront *Herr* Hitler in the defense of Poland."

"What about the submarines?" asked 'C.'

"Still in port according to our observers. However, they appear to be fully supplied and probably ready to sail in short order."

"As you indicated last week, the deployment of the submarines will probably give us a good clue of time."

"Surely, but even their current state suggests the date is near. You do not load food stores aboard a submarine just to have it sit there."

"Last month, you mentioned the *Deutschland* also put to sea."

"We lost her in bad weather south of Iceland. Both capital ships have maintained radio silence, so we have been unable to radio triangulate her."

"Also, not a good sign," added Menzies.

"Yes. Also, the day before yesterday, the Air Ministry and Fighter

Command tracked a large target with RDF along the length of the east coast. Would you believe the Germans sent out the old Graf Zepplein to reconnoiter our defenses?"

"The dirigible?" asked Menzies.

"Yes."

"What on earth for? It would certainly be no match for our fighters."

"No, it would not, but as of the moment, we are not at war, and it did not threaten the coast . . . just tracked along most of the east coast. The Air Ministry fellows believe they were gathering electronic information about our Chain Home RDF system."

"You mentioned disturbing news on the telephone," 'C' stated with a touch of impatience.

Pike cleared his throat and looked both men in their attentive eyes. "Several days ago, the exiled German physicist, Professor Albert Einstein sent a personal letter to President Roosevelt offering a chilling and candid assessment of German scientific work to develop an atomic explosive."

"Atomic explosive?" asked Menzies.

"As I understand the theory," interjected 'C,' "if a specific heavy atom of refined uranium, I believe, is somehow split, it will release an enormous amount of energy, but more importantly causes other adjacent atoms to split in what the scientists call a chain reaction. If possible, the explosive power would be many more times greater than any other explosives known to man."

"Apparently, Professor Einstein believes it is possible."

"Do we know what the American president intends to do?" asked 'C.'

"No."

"Does Winston know of this?" Sir Hugh continued.

"Yes. He hosted Einstein at Chartwell as the professor left Germany for America. He invited me to join them. The theory was discussed. However, our impression was the practical translation was more distant than this letter indicates."

"Do you have a copy?"

"No."

"Do you think you can obtain a copy?"

"No. Perhaps Winston can through either Einstein or Roosevelt. He is a friend to both."

Admiral Sinclair looked out the window as he considered the information. He glanced at Menzies, and then looked to Admiral Pike. "I would suggest you brief Winston as soon as possible. We need to develop the details on this work. We are at serious disadvantage today. We cannot allow the Germans to extend their advantage with this new explosive no matter how remote the possibility may be."

"I will do so at his earliest convenience. I shall ask him to make the approach to Einstein or Roosevelt if he agrees. Anything from Station X?" Pike asked in reference to the Government Code and Cipher School at Bletchley Park and the newly acquired Enigma equipment.

"Nothing. Denniston is working on gathering the best minds, but as he told us, the fourth rotor exceeded our collective ability."

Admiral Pike nodded his head in reluctant recognition. The meeting concluded without ceremony or smiles. Each of them knew how grave the situation appeared. They also knew quite well it was the burden of the professional intelligence officer.

—

Saturday, 5.August.1939
Westminster, London, England

THE journey from Hawarden, Chester and Manchester took the better part of the day. Pilot Candidate Brian Drummond was feeling more confident moving around the country. The room at the Savoy Hotel on the Strand was waiting for him as the note from Group Captain Spencer indicated. Although it was not the first thing he wanted to do, he knew it was the proper thing to do. He needed the small notebook he always carried to help him dial the right number.

"Spencer's."

"Group Captain Spencer?"

"No sir. Is this Mister Drummond per chance?"

"Yes, it is."

"One moment, sir. Group Captain Spencer has been waiting for your call."

"Brian."

"Yes, sir."

"Pleasant journey, I trust."

"Yes, sir. Long but no problems."

"Good. Do you have any plans for supper?"

The question stopped Brian cold. Virtually his entire conscious thought had been focused on one thing – Miss Anne Booth. Although he had not talked to her in a week, he wanted to see her and could only hope he would be able to link up with her.

The hesitation told John what he wanted to know. "I am sorry, Brian. You have plans."

"Well, sir. To be honest, I was hoping to see a girl I met."

"I see. I quite understand. If you have no plans tomorrow, I have us a date to see my uncle."

"Mister Churchill?" shouted Brian hopeful it was true.

"Yes, Mister Churchill."

"Great Scot!" he exclaimed, surprising himself with his choice of words.

A short laugh punctuated the conversation. "It seems you are picking up the idiosyncrasies of the mother tongue rather quickly, Brian."

"It just sorta popped out, sir."

"Quite all right, actually."

"When and where do I meet you or him, or whatever you've arranged, sir?"

"I will pick you up at the hotel, say half eleven. We will drive out to Chartwell for lunch."

"Fantastic. I'll be ready."

"Good," said John with a chuckle. "Have an enjoyable evening Brian. I will see you tomorrow."

"Thank you, sir. Good day."

Brian shouted and jumped around the room like a little kid who just found his most desired present underneath the Christmas tree. He was actually going to meet the man Malcolm Bainbridge said many times was one of the greatest leaders of all time. Wow, would he have something to tell Malcolm, he kept telling himself.

The euphoria about tomorrow passed into the 'think-about-it-later' category as the image of Anne Booth, and her earthly delights, returned to him. He took his jacket, shirt and necktie off to wash his face. Refreshed, he sat by the telephone, took a couple of deep breaths and dialed her number.

"Hello," Anne said in her soft, delicate, seductive voice.

"Anne?"

"Yes."

"This is Brian."

A moment's hesitation did not make him feel very good or encouraged.

"Brian, how sweet. Where are you?"

"London."

"At the Savoy?"

"Yes."

"What are you in town for?"

Brian did not want to appear too obvious. "My friend, Group Captain Spencer, invited me to London to meet Mister Churchill."

"Really?"

"I get to meet him tomorrow. Isn't that great?"

"I suppose so."

"Aren't you impressed?" asked Brian as he wondered why she might not like Mister Churchill.

"Certainly," she answered without conviction.

"I really want to see you, Anne."

"Brian, I have a commitment tonight."

The words stabbed him directly in the heart. His shoulders sank with the disappointment. Anne must have sensed the feelings. "Let me see if I can change the commitment," she said. "I will call you back. You are at the Savoy, correct?"

"Yes."

"Don't move."

Brian felt a little foolish. She was, after all, older than he was. She probably had a date with someone else. It had been a month since he had seen her, and he had only talked to her once on the telephone. What was he going to do if he did not see her? He had already told Group Captain Spencer he had a date. He would just have to spend the night alone.

The telephone rang. It was Anne. She had changed her commitment and asked him to come out to her flat. He remembered most of the directions, but asked for a repeat just to make sure. There would still be most of the evening available by the time he arrived. His heart was beating faster as he hung up the handset.

The Underground swept him through the series of subterranean tubes to Knightsbridge Station. A short walk passed the famous and luminous Harrods department store placed him in front of her door – No.14 Beauchamp Place. Brian stood looking at the polished brass knocker on the black door like it was about to spring at him. Before he could move, the door opened. Anne stood there in a bright, flowery dress with an expression of expectation. As he ascended the eight steps to the door, Anne extended her arms to embrace him. The feel of her body quickened his heart as it had done before. A long, full kiss accentuated the embrace.

"It is so good to see you again, Brian. I wasn't sure if I ever would."

"It's great to see you, Anne."

"Come in."

The interior of Anne's flat sparked a clear image of the night he spent with her. He wanted to feel the same sensations, the enthusiasm, and the energy he had felt with her that night a month ago. The bright smile and sparkle in her eye gave him the impression she felt the same

Brian could not determine especially what it was about Anne that made him feel so good. She fed him, made sure he was not thirsty, kept the conversation on him and generally gave him her undivided attention. Brian tried several times to learn more about Anne, but she skillfully deflected his queries. He accepted the path of the conversation although his curiosity about the magnetic woman near him kept a certain level of anxiety bubbling until her smile disappeared, her eyes dropped and a serious aura enveloped her.

"Brian, there is something I must tell you."

"You have a boyfriend," said Brian acknowledging what was on his mind.

"In a manner of speaking." Anne stopped short of saying the words since she could not find an easy way to say it.

"What does that mean?"

Anne decided the quick, short and to the point approach was best. "Brian, I am a courtesan."

"What's that?"

He was even more innocent than she thought. "A prostitute. I stay with men for money."

"Jesus Christ," Brian shouted standing up immediately. "I know what a prostitute is." Disbelief washed over him. He smiled. "You're kidding?"

"No, I'm afraid I'm not. I'm terribly sorry, Brian."

"If you're a prostitute, why haven't you asked me for money?"

"Jeremy paid me."

The shock sawed through Brian. "Damn him. Why did he do that? What did I do to hurt him?"

"Brian, he likes you. He thought he was doing a good thing for you, something you would enjoy."

"Well, I didn't," he lied.

The lie hurt Anne's feelings even though she knew his reaction was normal under the circumstances. She felt bad for him. She wanted to hold him and make him feel better.

"How much did Jeremy pay you?"

"I really do not think that is important."

"I do," Brian shouted.

Anne did not like his anger. She debated whether she should answer his question or not. In the end, she decided the only way she might be able to have a relationship with him would have to be based on the truth. "A hundred quid."

"My God."

They sat in agonizing silence, neither one of them able to find the words. The distance between them grew. Brian stood to leave.

"Brian, don't go."

"I'm not wealthy like Jeremy," snapped Brian. "I can't afford you."

"I deserved that, but I would like you to stay for me . . . and because you want to."

Brian stood looking out the window with his back to her. His pride was hurt. He felt tricked and denied a friendship he thought he wanted. The carnal desire within him wanted her, but the realization she had been with other men – probably many other men – nearly extinguished the flame.

A summer evening shower poured down outside. There was no point leaving just to get soaking wet. What would it hurt anyway? His thoughts returned to money.

"I haven't got very much money."

"Brian, listen to me because I will only say it once. I like you very much. I would like to be friends. Money is not required between friends."

"You've been with other men."

"Yes, I have, and I probably will in the future as well, but that does not have to affect our friendship."

The thoughts rolling through Brian's head touched many nerves, sensitivities and rational sensibilities shaking him like a young tree in the wind. The new terrain of personal relationships stretched Brian dramatically beyond anything he knew or was even aware of. He had no foundation from which to judge the possibilities Anne was suggesting. The attraction kept him pondering the choices.

"If we base our relationship on trust and honesty, we can enjoy each other. I do not want you to be a customer, Brian, but I must make a living," Anne volunteered. She waited for some response from Brian. With the surprises of this topic of discussion, Anne's empathy for the emotional turmoil undoubtedly within Brian allowed her the patience the moment required. Brian's personal struggle gave him a defeated image. "Brian, we can have fun and enjoy each other. I will promise not to let my profession get in the way of our relationship."

"Why do you do this?"

"I enjoy being around powerful men. Most of the time they just want an attractive woman to accompany them to an event or to chat. And, to be honest and completely candid, I enjoy sex although that is unimportant to what I do."

"I don't know, Anne. I like you, but I thought we would have a normal relationship."

"What is normal?"

"I don't know."

Anne considered his words. "Do you have a girlfriend back in America?"

The straightforward question caught him by surprise and rather crosswise with his own emotions. The question pointed directly at the duplicity of the situation. There was no purpose to be served by his continuing affront. "Yes."

"Well, then, there you have it. I am not asking you to give up your relationship with your girlfriend, or any other woman for that matter. I am not asking for any commitment. I just want to be friends." She made sense to Brian, and he wanted to be around her.

"OK."

"Are you sure?"

"Yes. I just like being with you." He felt a little silly saying what he was thinking.

"Good. Then, why don't we go have some fun?"

"Sure," said Brian without really knowing what she wanted to do, or more importantly what he should do.

Anne Booth stood, grasped his hand and led him upstairs. Events proceeded without words. The caring between them contributed to the intensifying feelings. Brian's readiness progressed faster and further than Anne. He started to move her toward the bed.

"No, wait. This time I want to show you a better way." Anne's physical excitement rose with the thought of teaching, of molding, the gorgeous hunk of clay standing before her. "A good lover considers the satisfaction of his mate while working to achieve his own pleasure. Making love is far more rewarding and enjoyable than just having sex."

Anne gave Brian a hands-on lesson about female anatomy and erogenous zones. With extra effort on her part, she helped him understand the techniques of pleasure. Although Brian was not significantly younger, they were worlds apart in the art of physical pleasure. The intimacy mixed with laughter, experimentation, enthusiasm and sheer enjoyment. Both lovers spent enormous energy to please each other.

They used the majority of the night to practice the skills and demonstrate their feelings for each other. The words between them took on a new tone that illuminated the powerful magnetism of the flesh. The laughter and gaiety brought both of them an enjoyment unique to each of them. The mutual exploration of his innocence made the freedom of human expression even sweeter than either had experienced.

They collapsed into each other's arms as the shades of dawn began to fill the window. The dampness and opulent scents of human pleasure enveloped them like a warm blanket.

—

Sunday, 6.August.1939
No. 14 Beauchamp Place
Chelsea, London, England

BRIAN woke with the morning summer sun fairly high in the sky. Anne smiled back at him as he perched on an elbow looking over every smooth curve of her body. The deep blue pools of her eyes, her parted full lips and mounds of her pillowy breasts kept him from moving from her side. A quick glance at the bedside clock told him he still had plenty of time before he had to meet Group Captain Spencer. The importance of the impending meeting with the famous British politician could not impinge upon the attraction of Anne Booth.

A new force drove his actions. The persuasiveness and strength of the indeterminate force amazed him. The need for more of her moved his free hand to cup her breast. Anne let him have his way.

"Good morning," Brian said softly with unintended emotion.

Anne stretched and twisted slowly as if to accentuate the gracious curves of her body for his enjoyment. "Good morning to you, Brian." She kissed him. "How long have you been watching me?"

"Oh, only a few minutes."

"I hope you like what you see."

"Without question," he answered as he gently fondled her breast.

The state of his arousal did not take long for Anne to notice. She liked his youthful vigor and decided to take full advantage of the moment. "I want you, Brian," she said as she pulled him toward her ready body.

Their lovemaking continued to full conclusion for both of them. Anne knew the young American pilot showed great promise as a considerate, caring and skillful lover. Her own commitment to realize the potential of this man made her pleasure all the more rich and full. There was a new and unique excitement with Brian she had not felt for a long time, if ever.

With the lull of their descent, Anne's thoughts jumped to Brian's appointment. She looked at the clock and calculated the time. "Shouldn't you be getting ready for your lunch date?"

Brian quickly checked the clock only to realize over an hour's time lapsed in their union. "Jeez-o-pete, I'd better run."

"Well, hop to it, then, mister," she said as she playfully pushed him from her bed.

Brian felt a loss as he dressed. He wanted the feelings he enjoyed with Anne to be perpetual, and yet he also moved into the light of reality that today he might shake hands with history. The mood of his thoughts moved like a relentless machine toward the focus of the coming meeting. Their good-bye embrace and kisses were long, warm and sincere as they had been yesterday evening. She watched him as he disappeared down the quiet city street.

The near cloudless summer sky brought many people out to absorb the warmth. It would be a great day, he could tell. The journey back to the hotel was uneventful and without delay. Brian took a quick bath, shaved, carefully brushed his teeth and pulled a fresh shirt from his kit bag. He was ready and waiting in the lobby when Group Captain Spencer arrived.

With the usual greetings and kind words dispensed with, they departed south across the Westminster Bridge over the River Thames. John Spencer drove them through the winding, turning, south side city streets. The conversations encompassed Brian's progress through advanced flight training, the characteristics both good and bad of the Gloster Gladiator and the colorful men who were his fellow candidates. Brian also took the opportunity, as Jonathan Kensington had suggested, raising the issue of assignment to a Spitfire squadron upon completion of his training. John Spencer made no commitment to help but understood the request.

Most interesting for Brian were Group Captain Spencer's guarded, rather pessimistic observations about the dwindling peace. Brian could tell he knew more about the situation than he was saying and also knew better than to ask the questions John would not be able to answer.

As the buildings of the city gave way to the fields and forests of rural Kent, John told Brian about Winston Churchill, his present political situation, the near total political and social isolation, and about Chartwell, the Churchill family home since 1922. John described Chartwell as Winston's sanctuary where he could insulate himself from the pressures of political life. Much to Brian's amazement, Winston was an erstwhile pilot who had great affection for the pilots of the RAF. He was also a painter of watercolors and a prolific author. Brian's questions consumed the time up to the point he saw a small sign pointing up a one-lane country road even narrower than the road they were on.

Sunday, 6.August.1939
Chartwell Manor
Westerham, Kent, England

THE sign simply said, Chartwell, with an arrow indicating the direction off to the left.

"Here we are," announced John as he turned into the narrow entry gate in the high brick fence.

The two-story, large brick building gave Brian a sense of importance. The bushes, trees and numerous flowers added a touch of elegance to the attractive house. No wonder he finds peace out here, Brian told himself. A well-dressed butler waited for them at the front door.

"So good to see you again, Mister Spencer."

"Always a joy to see you, Mister Smithfield."

"This must be the Mister Drummond you have spoken so highly of."

"Quite right."

"How do you do, sir," said David Smithfield, the middle-aged Churchill family butler.

"Very good, thank you," Brian answered extending his right hand. The gesture was not expected although Smithfield did shake the hand that was offered to him.

"Gentlemen, Mister Churchill is waiting for you." They followed him down a narrow corridor and into a long rectangular room with large windows on two sides and a grand, clear view of the Kent countryside to the south and east.

The uncle and nephew shook hands, traded friendly jabs and turned to Brian. John introduced Brian to the renown, some would say notorious, politician, statesman, pilot, author, painter and Englishman.

"Brian, I have the pleasure to introduce you to my uncle, Winston Churchill."

The young American's pulse rate reflected his excitement. The thinning nearly bald, round faced, slightly stooped man took several steps toward him and extended his hand. An image flashed into Brian's mind. Churchill's appearance, demeanor and gait reminded him of a large English, bulldog. His dark blue bowtie with white poke-a-dots dominated his attire.

"Winston, this is Brian Drummond."

"Yes, yes," the venerable old man said as if he were swatting a fly. Churchill grasped Brian's right hand firmly between both his hands. "It is an honor to meet you, my boy. I might add, Missus Churchill, whom would otherwise enjoy the pleasure of your company, is on the Continent tending to a friend. I offer her regrets."

Brian could feel himself flush with embarrassment. How could an accomplished, world famous man be honored by an American teenager? Brian stammered with a response that did not make sense.

"That is quite all right, young man. I have had the same affliction all my life."

Brian was even more embarrassed. Churchill laughed in a friendly, grandfatherly manner in recognition of the impact of his humility. He motioned to the chairs.

"Shall we have some tea before dinner?" With their acknowledgment, Winston nodded his head to the butler who disappeared to fulfill the request. "My nephew may not have told you, Brian, but we are related."

The constant barrage of unexpected amazement kept Brian very unsteady and unsure of what to do. "What do you mean, sir?"

A slight chuckle and a glance to John Spencer preceded the pronouncement. "I am half American, did you know?"

"Yes, sir. Mister and Missus Spencer told me."

"Yes, my dear departed mother, the former Jennie Jerome, came from New York. So, my boy, I have always felt a special affinity with America. We are countrymen in a way."

"I didn't think of it that way," Brian said with no other words available to him.

"But, enough about me. I have waited patiently to meet you ever since John told me about you. If John's information is correct, which I am most certain it is, you are the first American citizen-pilot to join us in the coming struggle."

"I suppose so."

"You probably have little appreciation for the significance of what you are about to do, but let me assure you, the sacrifice and contribution will not go unnoticed. We are dreadfully short of fighter aircraft, and to an even greater extent, pilots to fly them. John tells me your flight instructor, a Mister Bainbridge, was it John?"

"Yes, sir. Malcolm Bainbridge."

"Ah yes, good English name, right, Malcolm Bainbridge, flew with us in the RFC during the Great War, with you in Four Three Squadron, the Fighting Cocks, was it John?"

"Perfectly correct."

Churchill's expression turned instantly solemn. "I pray to God above, this shall be the last time we shall have to call on young men to defend our little island."

Brian could only muster up a nod in acknowledgment. The weight of the old man's words nearly buried him.

"In many ways, as an Englishman, I must apologize to you that this situation has come to us. War with Germany is close. We shall not know many more days of peace."

The choice of the word, days, as opposed to weeks, months or years, instantly stood out in Brian's mind as the most significant among Churchill's grand words. Days! Derek Langston had been right. Brian immediately recognized he would not be through advanced flight training in time. Many thoughts came to him with that realization. Would he finish in time to participate at all, or would the fight be over before he could complete the course? Maybe he had gone to all this effort, only to be faced with returning home without accomplishment.

Churchill's face brightened somewhat. "John tells me you are a rather good pilot."

"I don't know about that, sir. I'm still in advanced training."

"A formality, my boy, a formality. You will be through with it in no time. By the way, did he tell you I have had my hand at flying years ago?"

"Yes, sir."

"I was never much good, but I did enjoy the experience. I never flew anything as impressive as you are flying. Gladiators, is it?"

"Yes, sir."

"Well, if you are as capable as my nephew indicates, you should be flying the new Spitfire fighter coming into service."

Brian's excitement shot through the roof. "That's all I've wanted since Malcolm told me about the new airplane."

"Then, so it must be. John, can you make sure our young American guest is able to realize his dream."

"I will do my best, uncle."

"If you have any difficulty, you let me know. I will see to it."

"You have far more important things to worry about. Brian and I have discussed it, and I do not think we shall have any problem with his posting."

"Good, very good." Churchill turned back to Brian. "You shall have your Spitfire, my boy. All we can ask is that you use the awesome instrument to its full capability and you keep the Nawzees from setting foot on this ground."

There was a sense of pleasure in his pronunciation of Nazis. It was like a slippery, slimy slurry sound more as an insult than a reference title. "I shall do my best."

"I am sure you will."

The conversation diverted momentarily from Brian Drummond to Group Captain Spencer's work at Headquarters, Fighter Command. Mister Churchill was thoroughly absorbed in the issues, concerns, positions, wants and needs of Fighter Command and the RAF, in general. The respect from both men for a man they referred to as, 'Stuffy,' was quite clear. It did not take long for Brian to ascertain the man they called 'Stuffy' was none other than Air Chief Marshal Dowding. They talked about him as the architect of the air defense, command and control system, the growth of the modern monoplane fighters, and something they called, 'RDF.' Brian's razor sharp curiosity and urge to ask a barrage of questions was almost uncontainable. He kept quiet, not wanting to disturb the conversation of the great man.

Without the slightest hint, Churchill turned to Brian. "What do you think of all this talk of war, young man?"

Pilot Candidate Drummond was stunned by the question. It was as if the renowned statesman was asking him, an eighteen-year-old Kansan, his opinion regarding world events. "Well, sir, I'd say if there's going to be war, then let's get on with it."

Both men laughed not at Brian's thought but at the audacity of it.

"Regrettably, my boy, I am afraid you shall soon have your wish, and I hope and pray you and your mates are up to the task."

"We will be, sir."

"I am certain," Winston said. His eyes lowered in thought just for a moment. "Some good pilots in Fighter Command," he mused. "I simply wonder if there are enough of you to stop the looming Nawzee avalanche."

"'Marshal Dowding's plan is the best we have," John added. "We have done well with the time since *München*, but we are still not ready."

"I know," was all Winston could say. An enormous weight sagged his shoulders.

The quiet, contemplative interval gave Brian the strength to ask his own question. "What do you think of Hitler?"

Winston Churchill looked directly into Brian's eyes as if he was searching for some tell-tale sign. "Per chance, have you read *Herr* Hitler's book, *Mein Kampf?*"

"No, sir. I can't read German."

"That's OK, neither can I, but we do have an accurate English translation."

Brian felt the flush of embarrassment wash over him, yet again.

"*Mein Kampf*, in German, means my fight or my struggle. The book is the philosophical meanderings of a disturbed mind. *Herr* Hitler is a monstrous product of the former wrongs and shame of the Versailles Treaty retribution after the Great War. He is a madman whom we have blindly allowed to garner power of the most cataclysmic magnitude."

"Do you really think he will start another war?" Brian asked.

"I suppose that is a question many people are asking. I believe he thinks we are impotent to stop him. Yes, I think we are irreversibly headed for war. Fortunately, we have finally drawn a line in the sand. Most unfortunately, I think *Herr* Hitler will ignore it, and we shall be at war."

A very solemn minute passed without a word or movement from any of them. The panoramic view of the English countryside drew Brian's eyes as he considered Churchill's grim view of the future. It was very difficult for Brian to ascertain whether the old man was a pessimist or a realist. He was definitely saying the same things Malcolm Bainbridge had been telling him for more than a year.

"John, you are closer to the sword's point, what do you think?"

"Well, uncle, the situation does not look good. Much to our dismay, your prophecy of nearly four years ago is coming to pass."

"No prophecy, John, just simple observation. Once one has read *Mein Kampf*, you know the source of *Herr* Hitler's madness and the extent of his ambition. His premise clearly states man is a fighting animal; therefore the nation, being a community of fighters, is a fighting unit. Mix in a dash of Teutonic pride and ardor along with their virulent image of Aryan supremacy, and *voilà* the Nawzee war machine poised to swallow and devastate Europe."

John Spencer turned to Brian. "Do not let him fool you. He has been the most studious, accurate and incisive of anyone in or out of government."

Winston lowered his head. "Yes, well, perhaps so. Unfortunately, I have been resoundingly unsuccessful in convincing anyone of the danger."

Smithfield entered the main room to announce supper was ready. The three men followed him into the dining room. The meal was exceptional although Brian barely took note. The conversation continued along political and military lines as if Winston was innocently curious about what his nephew had been up to since their last meeting. As time and questions passed, Brian's opinion changed to reflect the subtle interrogation as if Churchill needed an additional source to corroborate several elements of intelligence data. Listening to the two men enthralled Brian as well as gave him a chill on a couple of occasions about what was coming. There was profound respect for the capabilities of the German *Luftwaffe* and *Wehrmacht*. The respect, not fear, was passed on to Pilot Candidate Brian Drummond, indelibly etched into his consciousness.

The midday meal lasted nearly an hour principally due to the protracted conversation. As if on cue and in recognition of the meetings, telephone calls and other correspondence that occupied his uncle's life, Group Captain Spencer said, "I am afraid we must be going, uncle. It is well into the afternoon and I know you must have many things to tend to. Brian must return to his training unit at RAF Hawarden."

"Nonsense."

"Thank you for the delightful dinner, uncle. I look forward to our next meeting."

"The pleasure was entirely mine." Winston turned to Brian. Grasping Brian's right hand between his hands, he said, "As I said earlier, Brian, it was an honor to meet you. I feel I must apologize for the unfortunate situation we have placed you in. I would like you to return when Missus Churchill is here."

"Yes, sir. Thank you very much for your gracious hospitality."

Winston held up his left hand for Brian to stop. "Your youthful vitality is quite admirable, and we are deeply grateful for it. But, listen to me, son, an old soldier, sailor and airman, what we are soon to embark on is a treacherous journey through the darkest night. You must have faith in yourself, the righteousness of our cause and the inevitable victory. We shall win with the hearts and spirits of our young men. As we say in the Navy, Brian, godspeed and following winds."

Brian was speechless with the words choking him unable to come out. In the end, all he could do was nod his head and struggle to contain an overwhelming urge to hug the old man.

The ride back into London bubbled with questions, comments, observations and general awe of the afternoon. John Spencer took great pleasure in the impact his uncle had on young Brian Drummond, his adopted protégé.

Westminster Palace dominated the view as they crossed the River Thames. "I had better take you directly to Euston Station."

The image of Anne Booth returned vividly to Brian's thoughts. The smell of her and their pleasures filled his imagination. He wondered, could he see her one more time before he left? "Yes, sir. It will be late by the time I get back to Hawarden. I'd better be on my way."

The remainder of the journey through the streets of London to Euston Station passed by without much notice as Brian considered only Anne. He remembered with perfect clarity the details of the previous night, and the unmitigated pleasure and fulfillment he enjoyed with her.

"Thank you, sir, for enabling me to meet Mister Churchill. I cannot tell you how grateful I am."

"Nonsense, you already have, Brian. I am glad you could. Take care and fly safe. Oh, Brian, don't forget, you want the London Midland and Scottish to Manchester."

"Thank you, sir. I've got it," Brian answered as John Spencer departed.

Chapter 10

> The terrible thing about the quest for truth
> is that you find it.
> -- Rémy de Gourmont

Tuesday, 22.August.1939
The Admiralty
Whitehall, London, England

"**W**ELL, then, it would appear the fuse is lit," said Vice Admiral Sir Geoffrey Pike as his briefing ended. The latest observations from their 'friends' in Bremerhaven and Wilhelmshaven arrived in staccato frequency over the last two days. "How much time?"

"A fortnight at most," answered the duty officer. "It will take three to five days for the submarines to be in the most effective position. Those are the limits as we see them. Anything less and the boats will be out of position. Anything more and they would have insufficient supplies for extended operations."

"Very well. Thank you, gentlemen. I think it prudent for you to return to the listener. I will brief the First Lord and First Sea Lord." The duty naval intelligence officer and his assistant took their cue, gathered their papers and returned to Room 40. Sir Hugh had not been entirely forthright. He also needed to notify 'C' at SIS as well as his friend and principal Member of Parliament in Commons but saw no need for such disclosure. The First Lord would brief the prime minister and the cabinet.

Sir Geoffrey decided to make his telephone calls first before he briefed his administrative leaders. The first call went to 'C.'

"Yes," came the impersonal greeting. The voice was not Sir Hugh, but he did recognize Stewart.

"Good morning, Stew. This is Geoffrey."

"Good morning, Admiral Pike."

"Is Sir Hugh not well?"

"To be honest, admiral, no he is not. He is bedridden. Our physicians are with him."

"Our prayers shall be with him."

"I will pass them along to Sir Hugh. You have something?"

"Yes. We have receive numerous, corroborated reports from Germany. A half dozen U-boats put to sea on Sunday and another dozen just yesterday from Wilhelmshaven and Bremerhaven."

"So, war is near."

"It would appear so. Our estimate is a fortnight at the outside."

"Poland."

"You have a better view," Sir Geoffrey stated in recognition of the body of information developed by SIS. "However, from everything I am aware of, yes, Poland . . . and soon."

"I will ensure Sir Hugh has this news as soon as it is reasonable to do so. Will the First Lord brief the ministers?"

"I shall discuss it with him, but I would presume so."

"I will consult Sir Hugh before I discuss our assessment with the prime minister."

"We begin the long, difficult journey into night. Based on this latest information, I shall recommend to the First Lord and First Sea Lord that we alert the fleet to go to war footing."

"Agreed. Anything else?"

"No, but I am certain we are about to become very busy."

"Talk to you later."

The telephone went dead. He found the number he needed and dialed the standard telephone.

The most important and, they all hoped, most unlikely contingency involved the politician they most admired, respected and felt a kindred relationship – the Right Honorable Winston Leonard Spencer Churchill, MP.

"Churchill residence," the butler announced.

"This is Admiral Pike for Mister Churchill."

"Sir, I am terribly sorry, but Mister Churchill is unavailable." Every one of Winston's friends knew the phrase really meant he had been up most of the night dictating a speech, portions of a new book, or an article for one periodical or another, and was still asleep. "May I take a message?"

"I am afraid I must ask you to wake him for a matter of the utmost importance."

The hesitation broadcast the conflict between the request he just received and the instructions from his employer. Sir Geoffrey waited patiently for the faithful servant to consider the possibilities.

"One moment, if you please, sir. I will see if Mister Churchill can be disturbed."

Several minutes passed with no detectable sound on the telephone line. The noise characteristic of an open connection was not present.

"Yes, 'Jumper,'" came the deep, muffled, lispy voice everyone recognized.

"My apologies for disturbing you, Winston." Pike paused for a response. There was none. He could feel the fatigue on the other end of the line. "As we agreed a month ago, I am calling regarding our situation. All the conditions have been met. I am about to brief the First Lord with my recommendation the fleet go to war alert."

"Dear God above. So we shall have war soon."

Sir Geoffrey did not respond. He instantly recognized the poignant, prophetic phrase from Winston's speech in response to the Munich Accord last fall. The response was also characteristic of the man he admired most. Concern for the fate of mankind without slightest thought about the greatest political risk he had ever taken.

The weight of the approaching reality must have sagged the shoulders of the great man. Winston knew the meaning of the '. . . all conditions met . . .' sentence. The Nazis were in the final stages of preparation for what would be an orchestrated, fictitious, unprovoked attack on Polish citizens. The official and very public positions of both France and the United Kingdom placed them squarely on a path of confrontation. The world was falling into the rapidly tightening swirl of the vortex. There was no way out now.

"Quite right, I'm afraid." The Admiral hesitated to consider whether he should say what he was thinking. There may not be a better time, he decided. "Winston, for what it is worth, Stewart and I are fully with you. We shall prevail."

"I know you will be successful. There may be no other effort of more importance to coming events. I know you are well aware of the significance. I simply needed to say it. We have always been good friends, 'Jumper.' I know everyone will do their best."

"I understand. We will keep you informed of our progress."

"Thank you. Godspeed." The connection was broken.

Admiral Pike replaced the handset, closed his eyes and cradled his face in his hands and elbows. The only thoughts filling his head encompassed the risks, the sequence of events soon to be carried out and the consequences of an error whether intentional or accidental. He also pictured Winston in his robe sitting in the big leather chair in the famous study at Chartwell doing probably much the same thing – contemplating the importance and consequences of failure.

Sir Geoffrey gathered his thoughts, pushed the small lever on his intercom box and asked his receptionist to gain him an audience with the First Lord – His Majesty's Government minister and senior, civil authority for the Royal Navy – and the First Sea Lord, the Navy's most senior admiral and leader. The reply came quickly. Pike rose from his desk and walked smartly out of his office and down the hall to carry out his duty.

Friday, 25.August.1939
House of Commons
Westminster, London, England

Winston Churchill recognized the significance of the huddle gathered around him in the far corner of the Member's Lounge. Duff Cooper and Anthony Eden were usually close by in important conversations. The unusual participants in this particular discussion were Clement Attlee, MP, Leader of the Labor Party, and Arthur Greenwood, MP, Attlee's deputy leader, as well as Sir Archibald Sinclair, MP, Leader of the Liberal Party. The mood in the Member's Lounge marked a change in focus for the people's representatives. Nearly everyone knew and understood they were less than a few days from war. Their efforts were no longer directed toward trying to prevent war. They were now concentrated upon readiness for the inevitable. This particular early afternoon discussion avoided the amenities.

"Do you think the Emergency Powers Defense Bill will enable the correct actions?" asked Greenwood.

"I believe so," answered Winston. "The government has already authorized the mobilization. The Admiralty is on war alert and has begun the process of merchant ship requisition based on yesterday's passage."

"Did I hear correctly that the Foreign Office issued the Polish territorial guarantee pledge this morning?" asked Attlee.

"I confirmed it myself," said Eden. "It was delivered to the Polish and German ambassadors at nine o'clock this morning. The essential content was given to the press at noon."

"Does it still look as if they are going to attack?" asked Sir Archibald.

"All the signs say yes," offered Winston, not wanting to get into the details of the latest intelligence briefing he received this morning. He knew the signs were unequivocal. The massive build-up of infantry, armor and aviation units all along the Polish frontier along with the deployment of capital ships and the most feared U-boats could only mean one thing. The increase in criminal activity in the German ethnic regions of Poland held precisely to the pattern established several years earlier. "The French have mobilized as well," he added.

"I suppose this is it then isn't it?"

"Winston," said Clement Attlee wanting to get everyone's attention, "in my discussion with Lord Halifax this morning, he indicated the Foreign Office was increasing its pressure on the Poles to make some concessions to ease the tension somewhat. With this damnable German-Soviet Non-Aggression Pact and the HMG guarantee, it may be their only hope, and ours as well, to avoid general warfare."

Cooper looked to Eden and interjected. "Have you seen the words of this Pact?"

"Yes, at least the version obtained by the Foreign Office," said Anthony Eden. "Most notable among the terms is their intention to desist from any act of violence, any aggressive action, and any attack on each other, either individually or jointly with other Powers. They have made their pact with the devil, and they will most probably reap the wind."

The gathering remained quiet for a long moment as Eden's words sunk in. The agreement between two historical, cultural and ideological adversaries baffled all of them. The announcement three days earlier had stunned, some might say shocked, the world and especially the British who looked to the weight of the Soviets to counter-balance the expansionist aims of Hitler's *Lebensraum*.

"Maybe the Poles could strike a deal with the Russians?" asked Attlee. He turned to Churchill. "With your connections on the Continent, is there anything you can do?"

Winston considered the idea and the request, but knew everything had already been done. "My friend, we have done all we can do. I have met with Count Warczynski, as well as several other friends both within and external to the Polish government, many times in the past year. He summarized the Polish attitude quite succinctly just yesterday. He said, 'With the Germans we risk losing our liberty; with the Russians our soul.' The Poles have drawn the line, and we have joined them. It is time we stand firm."

"It means general war," said Sir Archibald.

"Yes, I am fairly certain it does mean general war, but I am not bright enough to see that we have any other choice given the alignment of the Russians and Germans as well as the posturing by Hitler and his henchmen."

"Perhaps it is not time to stand firm, but time to stand and deliver in the terms of the highwayman," observed Greenwood.

They all laughed at the image offered by their colleague. While Winston was confident none of these influential men believed they were helpless victims of a proverbial robbery or extortion, it seemed to summarize the feelings they shared.

"What do you know about all these rumors concerning atomic weapons?" asked Greenwood.

The group turned to Churchill. Several of them knew about Winston's involvement in this issue. He used all the sources available to him including his friend, Albert Einstein, and others on the Continent and in America. The topic was not a new one although the specter of general warfare with Germany made the possibility quite a bit more onerous. None of them believed it could be true, but none of them could ignore issue either.

Winston looked into the expecting eyes of each man around him. "It is my firm belief we are dealing with scientific hypothesis and political conjecture. There is no danger that this discovery, however great its scientific interest and perhaps ultimately its practical importance, will lead to results capable of being put into operation on a large scale for several years."

"But, what if you and your experts are wrong?" Sir Archibald asked the ultimate question of doubt.

"If I am wrong, then we shall all know a far darker world. However, I believe there are four compelling reasons supporting my position. First, the best authorities indicate a minor constituent of uranium is required and the extraction process is laborious and time consuming. Second, the chain reaction process, as the scientists call it, requires a large mass that must be very precisely exploded. It is not a simple explosive process. Third, any experiments would have to be accomplished on a large scale that would be detectable. And fourthly, the Germans have access to only comparatively small amounts of uranium in Czechoslovakia." Winston paused to let the information soak in and germinate. "For all these reasons, the fear that this new theory has provided the Nawzees with some sinister, new, secret explosive with which to destroy their enemies is clearly without foundation. Dark hints will no doubt be dropped and terrifying whispers will be assiduously circulated, but it is hoped that nobody will be taken in by them."

"Dear God above," whispered Greenwood.

"Quite," added Sir Archibald.

"Perhaps it would be a good idea to evacuate the children after all," Attlee offered.

"It is the correct action regardless. London will be a certain, ripe, irresistible target for the Germans no matter what weapons they may use," responded Churchill. "I must caution all of us to avoid any alarmist rhetoric. We must struggle with these dire issues alone. The evacuation of our children is grave enough without adding the supposition of secret atomic weapons."

"Agreed," Attlee acknowledged immediately invoking head nods from the others.

"We obviously missed the call," Winston said looking around the rapidly emptying Lounge. "We best move along. We most certainly would not want to miss any of this debate."

The others agreed. They stood and moved silently toward the Commons Chamber. Churchill silently thanked the others as well as their ancestors for the cool, calm approach to ominous situations that characterized the British culture and heritage. He knew this would soon prove to be one of those historic moments. He felt the warmth of the growing dawn as he began to emerge from some of his darkest years in both his personal and political life. His time was soon to be upon them. He waited with patient anticipation.

Saturday, 26.August.1939
Westminster, London, England

 THE intervening three weeks since Brian's last visit to London enabled him to make good progress with his flight training. The extra effort including the last two weekends gave him a whole two days. A letter and two telephone calls set a meeting with Jeremy Morrison at Shepherd's Pub in this early Saturday afternoon. Brian knew he had some things to say to Flight Lieutenant Morrison. After talking to him and speaking his peace, a call to Anne would help. He hoped to see her although several attempts to call her yielded nothing.

 As the railway coach swayed down the tracks, Brian extracted the two letters reading Jeremy's first.

> 14 August 1939
> Dear Brian,
> I understand I have an apology to make. I would like to meet you in London as soon as you are able to get a weekend pass. I have finally been transferred to a line squadron, No.74 Squadron, Spitfires, at RAF Hornchurch. Please call me at Chigwell 0765 at your earliest convenience.
> Jeremy

 Jeremy sounded sincere and concerned about their friendship. At first, Brian had not wanted to meet Flight Lieutenant Lord Jeremy Morrison, Esq. By the second telephone conversation, Brian began to feel less angry and more curious regarding Jeremy's friendship with him. Then, his thoughts turned to Anne. He reopened her letter.

> 6th August 1939
>
> My Dearest Brian,
>
> I want you to know you are very important to me and our relationship is special. I am enormously regretful we met under the circumstances we did, but I am immensely grateful for our meeting nonetheless.
>
> I am deeply sorry Jeremy's good intentions caused you such pain. I will speak plainly when I say what occurred between us has nothing to do with Jeremy or my profession. I want to show you in every way how much you mean to me. You are a very special person, Brian.
>
> I would very much like to see you again as soon as you are able, so that I may show in the most special way I know how. We have a good spirit between us, Brian. I want it to grow. Please call me the next time you are in London, or you can see me anywhere you would like.
>
> It has been a very long time since I've said these words, but
>
> I love you!
>
> with my warmest, wettest love,
>
> Anne

Her profession still bothered him although the attraction of her personality and her body overcame any resistance. Miss Anne Booth was indeed special to Brian. He needed to tell her when he saw her, hopefully tonight as they had discussed a few days ago. Brian believed what she said to him and in her letters. He wanted to believe her.

The journey from Hawarden to London made the time pass quickly. The conversations with a wide variety of British citizens from elderly men to young school girls kept Brian's mind off the coming events of the day, the situation in Europe, and the pride everyone felt toward men and women in uniform. Genuine pride from those around him made other concerns dissolve to inconsequential.

The characteristic stale smell of Shepherd's Pub greeted Brian although the appearance was noticeably different. The usual sea of RAF blue ebbed to a mere fraction of its common level. Jeremy sat at the peripheral table with a half empty pint of dark beer in front of him. Brian had four paces toward him when his lordship turned to recognize his former pupil. Jeremy stood with a smile on his face.

"Good to see you again, Brian."

"Good to see you, Jeremy," said Brian with a noticeably subdued tone.

"Why so glum?"

Brian wanted to shout at him about linking him up with a prostitute, but he was also thankful he met Anne Booth. He decided to wait until the subject came up. Jeremy's letter acknowledged what he thought he knew. Anne probably told him or Virginia. "Oh nothing," he said finally.

"How is your training going?"

"All right I suppose."

"What are you flying?"

"Gladiators."

"Pretty good aeroplane, don't you think?"

"It is certainly the most powerful I have ever flown and it handles fairly well."

"Well, as I said in my letter," he paused for the waitress to deliver a pint for Brian, "I have been posted to Seven Four Squadron at Hornchurch."

"You said, Spitfires, right?"

"Quite right, actually. Spitfires."

In a flash, Brian forgot everything else. "Spitfires, wow. What are they like?"

"Everything you have dreamed. One hell of an impressive machine, I'd say."

Brian's thought turned inward. "I've got to find a way to fly Spitfires."

"The right thing will happen."

"How long have you been at Hornchurch? Wait, first, where is Hornchurch?"

Jeremy laughed at the barely contained enthusiasm. "You still need work on your geography. Hornchurch is in the Northeastern part of London north of the River Thames. I have been there about three weeks."

"That's great."

"Yes, it is. So close to all the beautiful women and flying the best fighter in the world."

Brian's thoughts returned to Anne Booth. He wanted to see her, but he felt some anger toward Jeremy. "Have you seen Miss North, was it?"

"Of course, I have. I shall see her tonight. I understand you and Anne have hit it off."

Brian's resistance evaporated. "Why did you pay her to be with me, damn you?" he shouted at Jeremy.

Morrison nervously looked about the pub. Everyone heard the strong words from the young pilot, and tried to pretend they had not heard them. "You do not need to raise your voice."

"I'm angry."

"I gathered that." Jeremy paused to let Brian expose any other emotions. Satisfied the volatile emotions were reasonably contained, he continued. "As I said in my letter, I owe you an apology. I am terribly sorry if my good intentions offended your sense of propriety. I thought she would be good for you, a good way to be introduced to the night life of London."

The young pilot stood sharply knocking over his chair. "I don't need anyone buying women for me."

"I have to ask you to lower your voice."

Brian looked around the pub to notice the uneasiness he had caused. The twinge of embarrassment did little to dampen the anger he felt. The calm, steady, unemotional expression on Jeremy's face along with the cool, penetrating blue eyes persuaded Brian to gather in his emotions. Slowly picking up the chair Brian sat back down and leaned forward to wait for Jeremy's response although he knew no question had been asked.

Morrison's motionless position changed as if his muscles were tired of holding the pose. "It was meant as a gesture of friendship. You weren't supposed to know. Anne is about the best there is, and I surmised she might be good for you."

"What makes you think I'm not experienced?"

"Your age for one thing."

The reality struck Brian at the same instant Anne's gorgeous face blossomed within his mind. He was thankful he had met Anne. He was thankful for Jeremy's consideration even if he did feel offended. Brian wanted to be with Anne. The thoughts of the delightful woman softened his indignation. "I really do like her."

"I am sure you do. So do I," Jeremy added, "but, not the same way, now. Anyway, I am surprised you made it."

"Why?"

"You do know they have called up all the reserves, and Fighter

Command is on full alert, don't you."

"Yes, two days ago."

"Right. I thought they might cancel your leaves to condense your training."

"They gave me this weekend at my request since I've flown the last two and apparently my instructors are satisfied with my progress."

"Good. You should take full advantage of the weekend. It would appear we may be reaching the end of our season of peace."

"Is your squadron on alert?"

"Yes."

"Then, why are you here?"

Jeremy laughed strongly. "Tit for tat, aye? Well, fair enough. The squadron is on alert as we speak. Two of the four flights are at readiness. The rest of us are supposed to get some relaxation on a regular basis. This is my time."

"Are you going to see Virginia?"

"My Lord, I hope so. At least, as I said earlier, that is the plan."

"Good. Good."

A moment of pensive thought passed between the two men. Each man without recognition considered the women. They both in their own thoughts wanted the comfort, pleasure and inner peace the women brought to them.

"Are you seeing Anne tonight?"

"I'd like to."

"Excellent. Then, what do you say we gather up the ladies and celebrate."

"Celebrate what?"

"Peace. Celebrate peace. We shan't have much more of it, I'm afraid."

"Let me call Anne."

"Smashing. Then, I'll call Virgin. We will show this crusty old village how to have a good time."

The calls took less than two minutes. Both women departed for Mayfair and Shepherd's Pub within moments in anticipation of an exciting evening with the men they found most stimulating.

The two RAF officers barely took the first swallows of their fresh beers when Virginia North entered the club to give Jeremy a most demonstrative hug and deep kiss. Even Brian received a kiss on the cheek. Anne joined them within minutes. Brian experienced a rush of warmth as he watched her walk toward them.

The drinks for Anne and Virginia had barely arrived and the beginnings of an evening's entertainment agreed to when a Metropolitan Police constable with his characteristic dark trousers, white shirt, dark tie and peaked Bobbie's hat with silver crest walked into the pub.

"May I have your attention, please." He paused only a moment to ensure everyone's mindfulness. "All members of the armed forces are ordered to return to their commands immediately. You are directed to return by the most direct means. Thank you."

"Any reason given?" asked a lone army major from the far end of the bar.

"None, sir. Simply the recall."

"Probably another bloody false alarm," the major said loudly as the constable left.

"I don't think so," Jeremy said more to himself than to the others.

"Oh, Jeremy, does this mean we are at war?" begged Virginia.

"Probably not, Virgin, but I suspect we are close."

"Oh, Jeremy."

"I know, love." Jeremy's broad smile returned as he turned to the others. "Looks like we will have to postpone our peace celebration. Do any of you need a lift?"

"No," answered Brian. Anne and Virginia never did answer his question.

The two men shook hands outside the pub. The summer light of the early evening made the time fade. Anne took Brian's arm to walk with him to the Green Park Underground Station.

"I am sorry we did not have more time, Brian."

"Probably not as sorry as I am."

"When do you think I will see you again?"

"I have no idea. I've never been in a war before, but if Jeremy is right I should get some time off occasionally."

"Promise me you will come to me. I need you. I need to feel you."

"I will if I can."

"Please, Brian. I need more of you." Anne looked deep into his eyes searching for the sign she yearned to see. She was not disappointed. "Even if you only have a short time, call me and I'll come to you. I've never been to Chester. I understand it is a beautiful city."

"It is. I'll do my best."

They kissed and held each other for the longest time trying to capture what could be the last moments between them. Numerous men and women, mostly older, objected to their public display of affection. A few celebrated the feelings between them. Brian disappeared into the network of tunnels beneath the city.

—

Monday, 28.August.1939
RAF Hawarden
Broughton, Clwyd, Wales

"When did you get in?" Pilot Candidate Brian Drummond asked his friend and classmate, Pilot Candidate Jonathan Kensington, as the two men prepared for the day's flight training activity.

"About one in the morning, I should think. You were sound asleep along with everyone else."

"Don't you think you're taking a bit of a chance coming in so late when we have to fly?"

"Actually, not particularly."

The two student pilots showered, shaved, brushed their teeth and dressed for the day. Brian still had some difficulty with the British need to be dressed properly, meaning, they had to wear a complete uniform, necktie, tunic and all, to which they added their flying kit.

As was usually the case, the student pilots joined the mess for breakfast early in the serving period of one hour. The instructor pilots and other support personnel most often ate later.

"What did you do this weekend?" asked Jonathan as they walked to the mess.

"London. I met Flight Lieutenant Morrison."

"How is he doing?"

"Flying Spitfires with Seven Four Squadron out of Hornchurch."

"You don't say."

"Seems to be pretty excited about the Spitfire. Somehow, Jonathan, I've got to get posted to a Spit squadron."

"Where there's a will, there's a way, mate."

"Well, I've got the will."

"Did you see Anne?"

"Yes, just as they announced the recall. Which reminds me . . . didn't you hear about the recall Saturday night?"

"No, actually. I was incommunicado with my girlfriend until Sunday evening," Jonathan answered with an enormous grin on his face.

"Did you get in any trouble for not making the recall?"

"A bit of a fractious lecture by the duty officer when I arrived, but nothing very dramatic."

They entered the dining room of the Officer's Mess and were promptly served by the stewards. The normal morning fare was not particularly different from what Brian Drummond got on the weekends in Kansas. It was good food, well prepared. Brian had never eaten so well or been treated with such respect. It made his occasional bout of homesickness easier to deal with.

"What do you think about this alert business?" asked Brian.

"Sounds like we are getting ready for a row."

"I wish we were further along in our training. If the war does start, we might miss it. It could be over before we get to operational squadrons."

"Not likely, I should think. I imagine we'll be in this one for a while."

"Maybe, but I want to get to a squadron as soon as possible."

"I will second that. I'm not particularly partial to this training business. I'd just as soon do the real thing."

One of the flight instructors entered the dining room. "Kensington, Drummond, get your lazy arses down to the ops hut. You've both got an early go, and you are late."

They quickly finished their eggs, toast and tea before grabbing their flight kit bag and walked fast to the flight line. They would barely have time to let their food settle before they would be in the air.

The early launch meant two long flights or three shorter flights for the day which in turn meant a long, fatiguing day's work in the air. This day involved various formation tasks including aerobatics. Brian and his instructor, Flying Officer George Esling, would fly with Jonathan and his instructor in a four ship flight through a prescribed series of maneuvers trading positions at various stages. Brian now appreciated the precision of formation flight and was on the edge of actual enjoyment.

They progressed rapidly through the syllabus. The last flight in mid-afternoon gave them their first formal instruction in advanced aerial gunnery. They fired their guns at a large red canvas banner towed behind a Lockheed Hudson. This was the fun part of flight training, and they enjoyed it. The worst part of the training was the rigidity with which the instructors expected them to fly the maneuvers. Brian found the techniques quite similar to those used in hunting pheasants, doves, turkeys or other birds. He quickly began to master the art of aerial gunnery. As they returned to RAF Hawarden, Brian could only think about the invigoration of seeing bullets hit his target. This was the first inklings of what he had been working for and he liked it.

The successful day of flying gave both Brian and Jonathan a natural excuse to have a pint before the evening meal. A quick shower to freshen up along with a clean shirt for the evening meal helped them get to the Officer's Mess bar.

"Jonathan," said Derek Langston as Brian and Jonathan entered the room to join three other members of D Flight. "Did you hear the news?" Jonathan shook his head. "The government has issued evacuation instructions for all school age children in London."

"What do you mean?"

"The government must know we are going to war, and they must know London is a prime target, so they have ordered all the children to be moved from the city into the country where they will be safe."

"You're kidding?"

"The hell I am. Me Mum called to find out what was happening. Her sister, me auntie, lives in London, and she has two wee ones who are being sent to a village north of Southampton."

"I wish he were," interjected Ian Milner. "When Derek got the news, I called my parents. My mother said the same thing. My family lives just south of Norwich and many families in the area have already been contacted to provide sanctuary for children from London."

"I would say this is war," added James Roland.

"And, here we are in training and nowhere near the action," said Brian.

"Me dad says not to worry, the action will come to us."

"Maybe, but we won't do much in the OTU. We've got to get to a squadron."

"I just want a piece of those damn Nazis," said James.

"Maybe we should request immediate transfers to fighter squadrons," Brian said.

"Yes, indeed."

Brian watched Flying Officer George Esling walk toward their table with a nearly empty pint in his hand. "We could not help hear this fanciful bit of saber rattling among the squirrelly clump of young chicks." All eyes turned to Esling although no one responded to the jab. "So, you want to get into the fight, do ya? Well, fighting Germans is not going to be as easy as some of you might think."

"What might you know about fighting Germans?" said James Roland in a barely audible voice.

The sarcastic question did not escape Esling's attention. With anger in his eyes and growing signs of intoxication, Esling responded, "My father fought the Huns in the Great War, you little bastard. He's made sure I know what to expect, and it ain't going to be a cake walk. So, I'd suggest you not be quite so eager to get into this row."

At that moment, one of the other instructor pilots and probably a friend grabbed Esling's arm against meager resistance to lead him out of the bar. Several minutes passed without words as each of the young pilots considered the admonition.

"Right then, who is ready for another?" asked James.

Derek and Ian indicated affirmatively. Brian excused himself without justification and Jonathan decided to join his friend. There was little discussion in the barracks building between the two friends. Brian made an attempt at study, but eventually gave up as he struggled with the realization war was close at hand. If they were evacuating children, the government probably thought the Germans would soon be bombing London. He wanted Anne Booth out of London as well. Brian thought about calling Anne to encourage her to leave, but Jonathan distracted him from further thought.

They both wondered whether they would be up to the challenge despite their words of confidence and aggressiveness. The stories given to Brian by Malcolm Bainbridge were probably not too different from those George Esling received from his father. Reality began to weigh heavily upon the braggart tones of untested youth.

Chapter 11

> War is nothing more than
> the continuation of politics by other means.
> -- Karl von Clausewitz

Friday, 1.September.1939
No.10 Downing Street
Whitehall, London, England

"Prime Minister," said the chief secretary as he tried to wake Neville Chamberlain. The task did not take long although he waited for the leader of His Majesty's Government to sit up on the edge of his bed with his feet on the floor. The years had taken their toll on the frail man.

"What is it?"

"We received a cable from Poland. The Germans initiated a general attack about 30 minutes ago at precisely five o'clock, continental time."

"So, it is finally over. There shall be no peace."

"It would appear so, Prime Minister."

"Have you summoned the Cabinet?"

"Yes. They will all be here at half past six."

"Good. Please call Sinclair and Pike. We will need a thorough intelligence briefing."

"Already done, sir."

"Excellent. Then, one more thing before I take my bath. I would like you to call Winston and ask him to attend as well."

"Sir?" the chief secretary responded. As a loyal ally of Neville Chamberlain's over the decades of his public service, he was keenly aware of the fundamental philosophical disagreements, putting it mildly, that existed between his leader and Winston Churchill.

"Yes. You did hear me correctly. I intend to return Winston to the Admiralty, if he will have the job. I can think of no one who commands the respect of the Navy or the rest of His Majesty's forces as Winston does. We must now all recognize that Winston has been correct all along regarding Hitler's intentions. It is time to stand against this tyranny, and we need Winston."

"As you wish, sir."

"Oh yes, so sorry. One more item, if you please. Please issue the war alert over my signature."

"Yes, sir."

Prime Minister Neville Chamberlain was left alone in his dimly lit bedroom to reflect upon the events that conspired against him and his once noble endeavor to preserve peace in Europe. He sagged under the weight and wondered what sad and onerous events lay ahead for his country, the community of Europe and the world, for that matter. There was no longer any doubt history would record his efforts of the last few years as a failed policy, an enormous gamble lost, and this day as a demarcation in world history. A cataclysmic conflict would soon be joined and the British people along with their French neighbors and allies must rise to the challenge.

The entire cabinet sat in muffled conversation when the Prime Minister entered the large cabinet chamber with its rich green walls, deep grains of the wooden baseboards and nearly full room, square, green leather covered, cabinet table. One distinguished MP and member of the Conservative Party sat along the wall. Also sitting in the periphery of chairs were the Director General of the Secret Intelligence Service, the Director of Naval Intelligence, the Chief of the Imperial General Staff, and the heads of each of the armed services. Prime Minister Neville Chamberlain sat heavily in the large chair at the head of the table.

The Prime Minister waited with all eyes upon him and exhaled audibly. "Gentlemen, as I know you are all aware, the armed forces of Germany invaded the sovereign territory of Poland at precisely five o'clock this morning. I have asked Sir Hugh and Sir Geoffrey to give us a thorough assessment of the situation in Poland, the North Sea and Baltic approaches, and the deployment of Nazi forces. Sir Hugh, if you please."

The intelligence briefing including the numerous questions from several ministers took nearly two hours. The subjects ranged from the details of the ground and air actions in Poland to German Navy deployments world-wide. Expected German actions over the next few days were discussed in detail. The public German reason for the invasion was to protect ethnic Germans from the violent, random thuggery in the Polish Corridor between Germany proper and the East Prussian enclave as well as the port city of Danzig. A few questions regarding the readiness of the British Expeditionary Force and the French Army as well as the Home Defense Force were asked and answered.

"There you have it, gentlemen" said Neville Chamberlain. "Now, according to our treaty with Poland, we are obliged to declare a state of war and initiate operations against Germany to provide assistance to Poland."

"Shouldn't we make another attempt to find a peaceful solution?" asked Foreign Secretary Lord Halifax.

Prophetically, Chamberlain looked directly into the silent and waiting eyes of Churchill who simply returned the look with expressionless eyes, and then looked back to his Foreign Secretary. "I am afraid, *Herr* Hitler has ruthlessly snatched the time for peace from us. It is now time to stand against his tyranny."

A minor but spirited debate ensued. The forces that brought Great Britain to this day still clinged to the belief peace could be regained through patience, accommodation and diplomacy despite the aggressiveness and nationalistic fervor of the Nazi war machine. The tide of opinion regarding diplomacy in Europe, and specifically with Hitler and Nazi Germany turned toward a more rigid position. Recognition of the change was clearly understood by the tone of the discussion. The Prime Minister encouraged the debate. Churchill resisted his instinctive urge to join the debate. He was not a member of the government and felt it inappropriate to participate in an internal cabinet debate.

Neville Chamberlain patiently allowed the arguments to continue until the reasoning began to repeat. "Gentlemen, if there are no objections," he paused to scan the entire room, "it is time for His Majesty's Government to stand for the principles of our heritage. We intend to issue along with the French Republic an ultimatum to the German Government for the immediate withdrawal of all German forces from the sovereign territory of Poland by noon, Sunday, the 3rd of September, or we shall be in a state of war."

"Do you really think that is necessary?" asked Lord Halifax.

"Yes, I do," Chamberlain said with conviction. "Are there any other objections?" Total silence cast an enormously heavy blanket over the occupants of the Cabinet Room. "Then, so it shall be. The Foreign Office is instructed to issue the appropriate ultimatum properly coordinated with the French as soon as possible, but no later than this evening. Are there any questions?"

The consequences of the actions they were about to take meant the world they had known would soon be relegated to an infinitely distant history. No one in the room could avoid the importance of this moment in the history of Europe and, a few knew, the World. There were no smiles, bright eyes or heads held high. Thoughts of modern warfare combined with vivid memories of the carnage of the Great War, the War to End All Wars, produced a particularly vile, gnawing, acidic brew within them all. The foreboding of the rapidly approaching future capped the grave cabinet session.

"The last item before we adjourn. I have asked Winston to join this Government as First Lord of the Admiralty." A few of the MPs clapped, most did not. "Winston, thank you for your acceptance. We look forward to taking full advantage of your experience for the endeavor before us."

In an uncharacteristic terseness, Winston responded, "Thank you, Prime Minister. It is an honor to serve once again."

Chamberlain simply nodded in recognition. "With that, gentlemen, we are adjourned. May God see us through this crisis."

The shuffling of papers, movement of chairs and lack of audible words filled the room as ministers departed. In the ante-room, several attendees congratulated Winston Churchill without the gaiety usually afforded cabinet appointments.

Friday, 1.September.1939
The Admiralty
Whitehall, London, England

CHURCHILL walked the distance from No.10 Downing Street to the Admiralty building with the First Sea Lord, Admiral of the Fleet Sir Alfred Dudley Pickman Rogers Pound, GCVO, KCB. Only a few words of cordiality passed between them as they walked. The admiral escorted the new and former First Lord to his old office. The journey through the halls of the Admiralty to the office of the First Lord took longer than normal as career civil servants, some few of whom worked with Churchill during his first service, passed greetings and best wishes to their new minister. People who were too young to work for him knew of him and took the opportunity to greet, congratulate and thank Churchill for what he brought to the Admiralty. Everyone smiled and seemed discernibly happy although the country and the Admiralty would soon face war. Once the two of them were alone, Admiral Pound handed Churchill a single piece of paper.

"I thought you should see this message, sent in the clear earlier this morning."

```
FROM: ADMIRALTY
TO: ALL COMMANDS
message
WINSTON IS BACK!
end of message
```

Admiral Pound noticed the watery shimmer in Churchill's eyes.

"Thank you," Winston said. "It is good to be back."

"It is good to have you back, sir. The Royal Navy and the country need your leadership."

"We have a monumental task ahead of us. I would like to talk to

Admiral Pike first, and then if you would be so kind to gather the principal staff. We do not have much time to prepare for what is before us."

"As you say. I shall ask Admiral Pike to come in straight away. I will organize the staff meeting for ten hundred hours, if that is agreeable."

"Certainly, that should do just fine."

Now First Lord of the Admiralty received a more personal, focused and detailed intelligence briefing from 'Jumper' Pike. In addition, Churchill absorbed the latest information regarding the status of the Enigma device that had yet to yield any substantive information. The question of security was answered with the fact that only twelve people were on the ULTRA access list which meant they knew of the existence of the little boxes locked in a safe in a guarded secure room at Station X, and they knew of the information eventually to be derived from the little boxes. The First Lord also asked very specific questions regarding the deployment of German capital ships and submarines. Churchill already knew the intimate details regarding the performance and capabilities of the German battlecruisers, new fast and quiet submarines as well as coastal defense gunboats.

"Do we have anything from ULTRA that might give us more insight into the rationale for Hitler's actions and what his real intentions are?" asked the First Lord.

"No, we do not. We have received an enormous volume of traffic at all levels although GCCS has not been able to find the right code wheel sequences to break the encrypted messages. In time, we shall know but not at present."

"Please let me know as soon as they have been able to read the material received plus or minus a few days from today."

"Yes, sir."

A knock on the door announced time for the requested staff meeting. The arrival of the additional staff officers at the appointed hour concluded the First Lord's personal intelligence briefing. Admiral Pike remained as the others entered. Each of the officers knew Winston Churchill from his reputation as a defense advocate, as a previous First Lord and as lone voice against Nazi Germany's explosive military expansion. Introductions were made by Admiral Pound for the benefit of the First Lord. After the cordiality, the leadership of the Royal Navy got down to business.

"It is my understanding from Admiral Pike, each of you are current relative to the situation in Poland as well as with German naval forces deployment worldwide." Churchill paused to confirm his understanding. "Excellent. Then, I am certain you will all agree the Nawzee response, if we receive any, will be to ignore our ultimatum." He again paused to receive the affirmation of the senior officers of the Royal Navy. "We must therefore use the remaining few days to prepare for general naval war with Germany. Let us use the time wisely to ensure we are prepared for proper defense, and I would appreciate your thoughts on offensive operations as soon as we can be ready."

For the next three hours, the assembled naval leadership discussed, argued, projected, planned and estimated the actions of the Royal Navy and the range of responses likely from the Germans. A broad consensus yielded the strategy for the early prosecution of the war. Every effort would be made to localize all the known submarines as well as the capital ships especially the DKM *Deutschland* and DKM *Graf Spee*. Initially, the Royal Navy would track and neutralize the German Navy rather than take on all-out offensive operations.

"With these agreements, Admiral Pound, please issue the appropriate orders, deploy our ships to support our plan and make all preparations for war with Germany. I intend to issue the execution order precisely at eleven hundred hours to match the Prime Minister's deadline. Let us be ready, gentlemen." Churchill saw nods around the room. "I believe our work is finished for now."

The officers did not need an additional cue. They departed promptly and quietly. The First Lord stood alone in the middle of his old office with his fists on his hips and his enormous, unlit, Havana cigar projecting straight out from his round face. The large wall maps of the immediate environs of Great Britain and the world were the same maps he ordered installed when he was First Lord the first time in 1915. The feel of the office refreshed his memory. The thrill of action so clearly an addictive elixir for Winston added the real spice to the memories of the First Lord's office. He opened a wooden casement covering a special map and much to his surprise saw the markers indicating the Royal Navy ship positions in 1916, the day he left the office after his first tenure as First Lord.

The remainder of the day consumed his attention as he established his routine, familiarized himself with the current tools of administration and most significantly the communications assets of command. Among the essence of office, Churchill took numerous telephone calls from friends, relatives and colleagues congratulating him on regaining his ministerial position. The excitement of being back in harness could not overcome the gravity of the situation in Europe.

Well into the evening, not quite alone in the building, Winston called Chartwell to his beloved companion, Clementine, to tell her about the day and the need for him to stay in town for the night. He then called his good friend, Bill Stephenson, to meet him at the Savoy Grill for supper.

Friday, 1.September.1939
Savoy Grill
Savoy Hotel
Strand
Westminster, London, England

THE walk in the warm, summer, evening air from the Admiralty building at the east end of Whitehall down the Strand gave Winston a welcome break. The mood of the city was surprisingly bright despite the news of the day. Winston could feel his energy level and enthusiasm rising as if he were receiving a transfusion from the public around him. It worked every time.

The Savoy Grill was not particularly well attended this evening. A table along the far wall waited for him. Bill Stephenson sat alone with only a barely touched glass of red wine in front of him. Stephenson stood as Winston approached. His friend and venerable statesman appeared markedly exhausted, drained, and void of color.

"Good evening, Bill," Churchill said softly.

"Good evening to you, Winston, and congratulations on your appointment."

"I am not sure whether it should be congratulations or condolences, actually."

"Are you all right, Winston?"

"Yes, yes," he responded waving his hand as if he were shooing a fly. "It has been a long day."

Stephenson motioned for the waiter. As was their custom, the two men knew exactly what they wanted, ordered and settled into what they both knew would be a working meal. The usual social conversation between friends filled the time not that they needed to use filler. They had to remain friends. First, their appetizer and then the delightfully displayed entrees were presented without ceremony. With the withdrawal of the waiter, the conversation changed.

"Bill, I am certain you are aware of events, so I do not need to tell you what will soon be upon us. His Majesty's Government along with the French Republic issued an ultimatum to the Nawzees for the immediate withdrawal of German forces from Poland. The deadline is midday, Sunday, the third. There is no expectation the Germans will respond to the message and certainly not favorably. The Government is making all preparations for war."

"As we would expect."

"Yes, quite right."

"Will Labor share in a national coalition government?" asked

Stephenson in his naturally curious manner.

"Apparently not. Neville did ask both Clement and Archie to join the government. From my private discussions with the leaders of the opposition, they refused outright without consultation. While they will not press their point, they both advocated their position that Neville should step aside in favor of my premiership. They indicated they would gladly join the government, if I were prime minister."

"It is only a matter of time, I should think."

"Perhaps, but this is no time for political rancor and maneuvering. We must pull together to the same stroke. Besides, I have given the Prime Minister my undivided loyalty and support. I will do nothing to undermine his position now that we are at war."

"The word on the street and among the press is, you are the best man for this crisis."

"I shall not succumb to the siren's song."

"The public mood is that of reality, not mythological seduction," Stephenson responded.

Winston chose to ignore the observations of his friend and change the subject. "I wanted to talk to you privately about a far more important task."

Stephenson knew what was coming. He kept silent letting Churchill control the pace of the conversation. The newly reinstated First Lord of the Admiralty casually looked around the room to satisfy himself of potential listeners. "I would like you to visit Washington to establish a personal and private communications link with President Roosevelt."

Again, Bill Stephenson recognized the logical questions, knew not to ask them, and took pride in the initiative, insight and strategy of his good friend.

"We shall need the ability to have frank, candid discussions with the American leadership even though the public position is one of staunch neutrality. I believe Roosevelt's private position does not match the policy he must represent. The Americans must be coaxed and cajoled to join us if we are to prevail. We must assist the President. Thanks to our late effort of rearmament, we are ill-prepared to take on the Nawzees."

"Agreed."

"Do you think you can gain direct access to the President?"

"I think I can. A personal letter from you would help."

"You shall have it first thing in the morning."

"Then, we should have no problem."

"Excellent. I shall wait for your successful return before taking the first step."

This time Stephenson looked around the room, then back to Churchill. "What should we do about our box?"

The First Lord took a characteristic cigar from his jacket pocket as he considered Stephenson's question. It would be a good faith move toward the American President. The intelligence community had a suspicion for some time the American cryptographers possessed their own version of the German device or possibly the Japanese equivalent. The desire to compare notes, to have the best information available, could not overcome the uncertainty of this initial contact. After all, he was simply the First Lord of the Admiralty, the equivalent of the United States Secretary of the Navy, not the Prime Minister, trying to talk directly to the President of the United States. Although he knew the contact was right, even he had to accept the possibility it was quite presumptuous and outside any form of protocol.

"We can do nothing at present. You must avoid any reference, explicit or implicit, to ULTRA, or any associated information or programs," Winston said quietly but firmly. "We cannot afford a mistake especially at this tenuous stage in our hoped-for relationship."

"Understood. When do you want me to leave?"

"Tonight, if you could, but tomorrow would be acceptable. Please let me know if you need any resources, or have any problems or obstacles."

"I will."

The remainder of the supper meeting covered the public information regarding the situation in Poland and the expected Allied response. Expectations in the media created their own spectrum of rumors, supposition and intrigue. The two men understood the need of the media to fill the information void, and gained comfort from the intimate knowledge they possessed. They both felt the urge to communicate the details, but cooler minds had to prevail for the moment.

Bill Stephenson glanced at his watch and grimaced upon the realization so much time had passed. "I really must be going, Winston. I am not able to keep the hours you do, and it would appear I have an arduous few days ahead. So, if you will excuse me, I shall be off."

"Quite all right, I should say. I wish you the best of luck on your mission. I know I do not need to tell you how important this journey will be, Bill. By the way, please tell Franklin, should you get to see him directly and privately, we shall refer to you by the code name of," Winston paused to consider the possibilities, "let's see. What would be appropriate?" Stephenson looked like he wanted to say something, but Winston beat him to it. "I think it should be, 'Intrepid.' That should do just fine. After all, you are an intrepid sole. Anyway, I would strongly recommend you act as our personal emissary for the moment."

"Very well. There should be no problem with any of that."

The two men shook hands. Bill Stephenson departed leaving Winston Churchill to finish his coffee and French pastry dessert.

―

Sunday, 3.September.1939
The Admiralty
Whitehall, London, England

"**M**Y apologies for disturbing your weekend and the Lord's Day as well at this very early hour, but I am certain each of you fully appreciates the need," the First Lord began. "We have received only a feeble statement from the German Ambassador in London regarding the peaceful intent and justifiable actions of the *Wehrmacht* in Poland. It is comforting to know the Nawzees mean no harm," he said with considerable sarcasm. General laughter greeted his words. Winston waited for the laughter to vanish. "We have not received any communication, formal or otherwise, from the German Foreign Ministry. It would appear we shall be at war with Germany in another few hours. Are we prepared?"

"Gentlemen, your reports, please," commanded the First Sea Lord.

Several staff officers conveyed the readiness reports from the various fleets and independent squadrons within the Royal Navy. In addition, the readiness of the Royal Air Force as well as the British Expeditionary Force in France were discussed relative to the Navy vessels assigned to coastal protection and Channel surveillance. The sailors and officers were ready.

Admiral Pike provided the assembled staff the latest information from the continent. The progress of German forces staggered the imagination of every military professional and tactical practitioner. Considerable discussion about the specter of modern armor warfare, which the Germans called *blitzkrieg* or lightning war, occupied the better part of an hour. The rate of advance on the ground and the devastation wrought by the *Luftwaffe* was almost beyond comprehension. Several members of the Naval Staff questioned the accuracy of the intelligence, Admiral Pike among them. While significant confusion kept accuracy an open issue, the preponderance of data corroborated the information. Within weeks, if not days, the *Wehrmacht* would be on the outskirts of Warsaw.

"Do we know the whereabouts of their capital ships and submarines?" asked Winston.

"The *Graf Spee* is confirmed in the South Atlantic although we are not certain of her precise location."

"Does she have escorts?" asked Churchill.

"No. The *Deutschland* is believed to be in the North Atlantic, but we have not seen her in several weeks. Nearly two dozen U-boats are confirmed at sea and perhaps more. We can only surmise they are spread across our shipping lanes and covering the approaches to the Home Islands."

"Then, every ship of the line must assume they will soon be confronted with a hostile submarine threat," said the First Lord.

"They cannot be everywhere," observed one of the assembled flag officers.

"Quite right, you know, but unless you can tell me where these submarines are, I respectfully submit, we cannot ascertain where the threat lies, and therefore, we must assume it is everywhere. Do you disagree?" asked Churchill more sharply than usual. The fatigue of long hours weighed heavily on each of them.

"No, sir."

Silence filled the room for more than a minute. No one moved or shuffled papers. Each man drew inward with contemplation. Winston Churchill looked at the large clock perched among the maps on the wall.

"Gentlemen, the time is now, 10:37 hours. His Majesty's Government issued our final ultimatum at 09:00 this morning. We have a third of an hour to issue instructions to the fleet. Is there any further discussion?"

No one responded with words. Every man in the room shook his head in a negative gesture. Churchill rose from the conference table to lift the telephone on his desk. He made two calls, asked the same question and received the same answer.

"I have once more confirmed we have received no indication the Germans are or intend to comply with the ultimatum issued by His Majesty's Government. In just a few minutes, we shall be at war with Germany. I would like this message issued to all commands, ships, units, and facilities of the Royal Navy precisely at 11.00 hours Greenwich Mean Time, noon in Berlin." Churchill handed Admiral Pound the single piece of paper.

The First Sea Lord took a quick glance at the note passed down his side of the table to his head of operations. "Make it so," he said with a confident, authoritative voice.

As the first tones of the hourly chime from the Clock Tower of Westminster Palace rang out across the city, the teletypes of the Royal Navy around the world clicked out the simple, uncoded, operational message.

```
FROM: FIRST LORD OF THE ADMIRALTY
TO: ALL ROYAL NAVY COMMANDS, UNITS AND FACILITIES
message
WE ARE AT WAR.   COMMENCE HOSTILITIES AGAINST
GERMANY.
end of message
```

Sunday, 3.September.1939
RAF Hawarden
Broughton, Clwyd, Wales

THE Sunday morning allowed each of the pilot candidates of D Flight to wake by some natural process. The overcast sky and drizzling rain added to the slowness of the morning. Brian Drummond and Ian Milner, the only two members of D Flight from farming communities, rose early as they usually did and completed the morning meal before most of the other pilots woke. The relaxed time and lack of disturbance kept each of the young men to themselves. Ian read a book as the others began to get dressed. Brian reread letters from Becky and his parents.

The loudspeaker shocked each of the fledgling pilots. "Attention, please, for an announcement from the Prime Minister."

Brian glanced at the clock on the center wall. It was 11:15. Several of the pilots looked at each other. This was the first time any of them heard a loudspeaker announcement from any one of importance beyond the confines of the airfield. "What the hell could be all this important to muddle up our only time off?" shouted Peter Barker in his angry Liverpublian accent.

Every member of D Flight and probably every member of the Royal Air Force turned their faces and attention to a funnel shaped device on a pole, in a building or hangar, that they commonly referred to as a Tannoy, or loudspeaker. The Tannoy blared out all the important announcements, alerts or other bits of information. This occasion was no different.

"This morning," came the tired voice of Prime Minister Chamberlain, "the British Ambassador in Berlin handed the German Government a final note stating that, unless we heard from them by eleven o'clock that they were prepared at once to withdraw their troops from Poland, a state of war would exist between us. I have to tell you now that no such undertaking has been received and that consequently this country is at war with Germany. May God see us through the struggle ahead."

"My God, he's done it," said Peter.

Several pilots including Brian wondered who 'he' was. Did the pronoun refer to Hitler or Chamberlain? Who was being blamed for starting the war?

"What do you mean?" asked Ian.

"Chamberlain. He has finally put his bollocks to it. We have us a row."

"We have been through this," commented Jonathan Kensington.

"Yes, we have, but nonetheless, we have a war. We have to get through this training."

"We will whip those damn Krauts in no time," announced Derek Langston.

"Maybe now that we are at war, they will let us finish early and get to a squadron," said Peter.

"Why don't we go down to the Ops building to see what they can do?"

"Right. Let's do it."

"They are going to tell you not to get your knickers in a bunch," interjected Jonathan.

"Maybe so, but it certainly would not hurt to try, now would it?"

"Maybe not, but don't get your hopes up."

The bravado continued as Peter and Derek got dressed. The rain, in part, along with a Sunday routine and cooler minds kept Jonathan, Ian and Brian in the residence building. Within ten minutes, the two enthusiasts were off to the Operations building.

The conversation quite naturally centered upon events as they imagined them to be happening. All the young men possessed were press reports about the German invasion of Poland. The lightning quickness of the air, armor and infantry onslaught shook their consciousness, thoughts and conjecture. What was going to happen next? Were the Germans going to attack Great Britain? Were they going to attack France? Where was Derek Langston's uncle, the BEF sergeant? What did he think of all this?

The questions created more questions, none of which had answers, although the opinions were ripe and flowery. The unanimity among the young pilots conveyed the enthusiasm, vigor and pretension with which they approached the crisis. The distance they enjoyed enabled their youthful swagger.

The characteristic rattling of rain upon the roof punctuated the conversation. Peter and Derek returned to shake off the wetness.

"What did they say?" asked Ian.

"The duty officer told us to talk to the chief instructor on Monday," answered Derek.

"Did he think it was possible?"

"I believe he gave us a good indication when he laughed at our question," Peter said.

The observation dampened their enthusiasm like the rain on the ground. Only the pitter-patter of the rain could be heard. The air in the room seemed heavy. Each of the young pilots felt the frustration of wanting to do something significant and not being allowed to join the process. They wanted their training to be over. Each of them in their own way knew they had sufficient training to fight the Germans although their instructors did not agree. Gradually, each of the men succumbed to the mood cast by the weather drifting off into their own activities. Brian reread his most recent letters and decided to write back to those he loved and cared about.

Wednesday, 6.September.1939
Fighter Command
Bentley Priory
Stanmore, Middlesex, England

THE first smears of dawn barely accentuated the broken clouds over Southeastern England. The incessant lights of London no longer illuminated the night sky south of the old monastery, now the headquarters of Fighter Command, as the declaration of war brought concerns over the possibility of night air raids. Group Captain John Spencer descended the dimly lit stairwell as he did every morning since the declaration of war. He liked to see what business might be at hand before he had to return to the chores of a command staff above ground.

The brain of the central nervous system of Fighter Command occupied a large room, simply called the Operations Room, in the recently completed underground bunker 50 feet beneath the surface of the picturesque countryside.

The enormous table that dominated the floor portrayed the whole of the British Isles as well as the French coastline across the Channel. The grid pattern divided the entire plotting board into coded squares that formed the basis of national geographic identification. Several young female RAF plotters sat casually around the board that had become their duty station. Less than a month ago, they began 24 hour operations support in the Operations and Filter Rooms that left them for the most part bored from inaction.

"Look alive, now," the chief plotter was heard to say.

The duty controller looked down from the observation deck surrounding three sides of the Operations Room. A block designating a hostile target near Mersea Island in the Thames Estuary appeared on the plotting board, the only target.

"What have we?" asked the controller.

"The Mersea Island searchlight battery along with the Observer Corps confirm a hostile target overhead heading west at an estimated altitude of Angels one two," came the reply.

The controller picked up the telephone and dialed the number for the Mersea Island searchlight battery commander. "Wing Commander Royston here. You have reported a hostile target," he stated.

"Yes, sir. We still have him. He's in and out of the clouds, but we still have him."

"What type of aircraft?"

"Twin engine bomber. We have identified him as a Junkers 88."

"You are certain?"

"Absolutely, sir."

"Very well, then. Keep on him. I'll send a few fighters to engage."

"We will track him as long as we can, but he is heading toward London."

"Understood. Thank you," Royston said as he hung up the phone.

The veteran controller lifted a green telephone which had no dialing ring.

"One-One Group," came the answer.

"We have a confirmed hostile over Mersea Island. You are cleared to engage."

"Very well," the group controller acknowledged as the line opened. He immediately picked up the phone to North Weald Sector Station. "Scramble, Green Section, One-Five-One Squadron."

Spencer visualized the sequence. Within seconds, three pilots would be running toward their Hurricane fighters. Within minutes, the three fighters would rise from the grass aerodrome of North Weald climbing at maximum power toward the East. The layers of broken clouds kept the Sun from their eyes until they were on top at about 8,000 feet. The black blotches ahead of them marked the airbursts from the Chatham Anti-Aircraft Battery. Somewhere among those ugly spots they knew a lone German bomber had been engaged. They continued to climb to the designated point in the sky called, the intercept perch. The point would be just beyond the range of the Chatham Anti-Aircraft guns and behind the intruder.

"Canewdon Chain Home reports three new contacts," announced one of the controllers as the plotters positioned a marker near Mersea Island.

Unfortunately, the still maturing early warning system including untested operators succumbed to a peculiar phenomenon associated with Radio

Direction Finding called back scatter. The RDF operators saw the fighters climbing to engage the enemy behind the large, stationary RDF antenna arrays and interpreted the reflected signal as German aircraft in front of them. The three fighters caused six more Hurricanes to launch which in turn caused more RDF returns and serious alarm within the command and control system.

Wing Commander Royston issued orders to notify Air Chief Marshal Dowding regarding a significant raid, certainly greater than the occasional reconnaissance flight seen previously, developing East of London, and then called the chief controller at No.11 Group. "Scramble Seven Four Squadron. Hostile raid East of London. Twenty plus aircraft."

"Very well." The chief controller then raised the appropriate telephone among the bank of telephones on the counter before him. "Scramble Seven Four Squadron."

—

As their colleagues from North Weald had done less than half an hour earlier, all twelve Spitfires of No.74 Squadron at Hornchurch aerodrome took off preparing for combat. Flight Lieutenant Morrison led C Flight toward their designated intercept perch.

The confusion over events in the sky above the Thames Estuary grew at a dramatic, frantic and cancerous rate. The confusion on the ground was serious. The confusion in the air was deadly.

The original aircraft that sparked the mêlée was a returning twin engine, Blenheim light bomber. The young Hurricane pilots set upon their hapless compatriot. Forty-seven bullet holes, one punctured fuel tank and one destroyed engine later, along with frantic calls from the injured crew, the fighter pilots recognized the mistake, unfortunately too late for themselves.

Seeing the attacking fighters, several of the Spitfire pilots of No.74 Squadron in the confusion and excitement mistook the Hurricanes of No.151 Squadron below them as German Messerschmitt Bf109 fighters and dove to engage. Jeremy Morrison checked his guns to ensure they were armed, pushed the throttle full open and rolled in turn to lead his two wingmen into the fight. As he closed with the twisting and turning fighters, a fortuitous combination of sunlight, cloud background and angles to the Hurricanes flashed the distinctive blue and red roundel wing insignia. At first, he noted the indicators with disbelief, and then heard his squadron leader broadcast over the radio the recognition error. He pulled his throttle back along with the stick moving the nose of his Spitfire and its eight guns away from the intended targets.

In the chaos of the first combat, the confusion of the engagement and the many conflicting voices, not all the Spitfire pilots made the connection.

Seeing themselves engaged by single engine fighters diving out of the Sun, the Hurricane pilots maneuvered to defend themselves.

During the twenty minutes of aerial combat resultant from war jitters, Observer Corps error, RDF back scatter and youthful enthusiasm, two Hurricane fighters were shot down. One pilot did not survive. Nine pilots were wounded and 17 other aircraft damaged to one degree or another.

While Jeremy Morrison did not realize the extent of the error as he landed his Spitfire at Hornchurch, the heavy cloak of regret, remorse and resentment sapped his body and mind. Before the engines were switched off, his squadron leader broadcast to all of them. "There will be no talk of this until we debrief, lads." Jeremy knew what that meant.

The ground crews were excited to hear the news of their first combat. The pilots drooped under the weight and said nothing as they made their way to the dispersal shack. Once inside, the emotions let go.

"Those were bloody Hurricanes," said Jeremy.

"How could they make such a mistake?" asked one of the young pilots.

"Enough of this lads," commanded the squadron leader. "We will put a cork in it until the intelligence blokes can sort their way through this." The pilots did as they were ordered, but continued the searching within the confines of their own thoughts. Each of them knew the seriousness of the mistake and wondered why.

The formal debriefing took the remainder of the day due to the nature of this particular engagement. The investigation by Fighter Command and the Royal Air Force would last nearly four weeks as they faced the reality of the errors and the implications to their delicate air defense system. They all knew in their hearts the combat in Poland would soon spill over into the rest of Europe and undoubtedly Great Britain. They all asked the question, could they correct the errors and be ready for the onslaught when it came? The question gnawed at everyone.

The early spikes of reality began to sink into the command structure. The war was less than three days old. The first combat test of the United Kingdom's untried air defense system succumbed to the flaws of immaturity. The breadth, reality and consequences of the failure would not imbue confidence in the British public and would also encourage the enemy.

The most biting wound of all came in the form of a joke. Within days of the fatal and damaging incident, references were whispered about the mistake. History would soon refer to the incident as the Battle of Barking Creek, a sarcastic reference to traditional theatrical portrayal of bumbling, comical buffoons stumbling through normal events of life. To those who had worked so hard to build the system, the reference brought the most pain.

Verbal orders issued the same day classified the incident over Mersea Island as a state secret, thus restricting all participants from discussing it publicly. The order also restricted the press from printing any element of the story regarding the so-called Battle of Barking Creek.

Chapter 12

*What we call experience
is often a dreadful list of ghastly mistakes.*
-- J. Chalmers Da Costa

Saturday, 16.September.1939
Westminster, London, England

The heavy, damp, smell of rain-soaked streets fit quite well with the dark, gray overcast. It was late afternoon, but it was also getting dark earlier, now. The thick fog began to develop adding to the isolation Brian felt. The warmth of summer was gone. Brian missed the warmth, even as he did in Kansas, as he walked the short distance from the Green Park Underground Station to meet Jeremy, Virginia and especially Anne at Shepherd's Pub. Anne had sounded so soft, warm and excited about their reunion. This time was different. He would introduce Anne Booth to his friend, Jonathan Kensington.

"It is certainly our good fortune we do not have to fly in this mush," said Jonathan.

"It sure is."

"It has been a dreadfully long time since I have been to London."

"Maybe, Jonathan, but I'm sure you know this city better than me."

"Actually, I have only been to London twice."

"Twice," barked Brian. "You mean I've been to London more than you?"

"I suppose so."

"Well, I'll be."

"Strange as it may seem, how many times have you been to Washington, DC?"

"None."

"There you have it."

Brian Drummond understood the point and let it pass. The spectrum of people passing them offered the distraction. The people fascinated Brian every time he visited the city. The vision of Anne soon filled his anticipation.

They walked into Shepherd's Pub 30 minutes early. The pall of cigarette, cigar and pipe smoke was unusually heavy and choking, Brian noted.

"Let's have a pint while we wait."

The two young pilots took several steps toward the bar.

"Aye, you blokes."

Both men turned to identify the caller. Jeremy Morrison sat alone at a table in the far corner. As they approached the table, a half empty bottle of scotch sat directly in front of him. The sunken posture and void expression of Jeremy struck Brian immediately. He glanced quickly to Jonathan who returned his look with questioning eyes. He did not stand.

"Flight Lieutenant Morrison, this is Jonathan Kensington." Brian held his hand toward Morrison.

"Good to see you, Brian. Nice to meet you, Jon," Jeremy said with solemnity and grimness.

"Let us get a pint," Brian said. The two young pilots returned with their glasses of the dark, English brew. Jeremy bolted another two fingers of scotch whiskey. "What's wrong, Jeremy?"

"What isn't wrong?"

Brian looked again to Jonathan who was just as puzzled as Brian. "What do you mean?"

"Ah, we are sworn to secrecy, now aren't we." Brian and Jonathan said nothing and did not move. "But, I have had enough liquor for me not to care, and I see no spies in this delightful establishment. Sit down and let me tell you what is wrong," he said with a growing slur to his words. "Oh, yes, I suppose you are already sitting, so let me continue."

"You don't have to tell us anything," Brian offered knowing that in his drunken state he might say something he would regret with sobriety.

"Nonsense. You have a right to know what you have gotten into, I should think."

Brian felt quite uncomfortable. He had not yet heard what was bothering Jeremy so much. He instinctively knew he did not really want to hear anything that dramatic. Jonathan shared the discomfort, and he barely knew Jeremy.

"Have you heard of our mortal combat of a week or so past?"

Both men shook their heads. They had only heard rumors about reconnaissance aircraft and attempts to shoot them down without success as far as they knew.

"You have not heard the saga of the Battle of Barking Creek?"

"No."

"Well, let me tell you. It seems our illustrious air defense system reported a lone German bomber inbound to London which turned out to be a solo Blenheim returning from a recce flight. Thinking our glorious capital city threatened, they scrambled a flight of Hurricanes. Then, for some God awful reason, they think the Hurricanes are 109s, and they scramble our entire frigging squadron of Spits. The short of it is, some of our young lads think they are really in a fight and they start shooting. Ah yes, you guessed it, this little *mêlée* results in two valuable Hurricane fighters shot down, one Blenheim light bomber seriously damaged, one pilot killed, nine pilots wounded and seventeen, count 'em, seventeen aircraft damaged. This, my young friends, is what the defense of these sacred isles rests upon."

Brian and Jonathan sat stunned and speechless. Jeremy bolted another finger of scotch.

"My God," Jonathan said finally.

"Yes, indeed. There wasn't one frigging German in the sky, nears we can tell."

"My God," he said again.

"We must pray to God the bloody Germans don't really come after us, or we shall all be speaking German, soon. *Sprechen sie Deutsche?*"

"What does that mean?"

"It means, do you speak German," interjected Jonathan.

Brian ignored the drunken, dark humor. "Is it really that bad?"

Jeremy laughed an acidic, angry chuckle. "One bloke died!" he shouted causing everyone in the club to look at him although he could not recognize or appreciate the situation. "Of course, it was that bad." His words become more slurred with each breath. He raised a finger to his mouth and whispered, "Oh, but we are not supposed to tell anyone."

"Maybe we should go?"

"Go where?" asked Jeremy with nearly unrecognizable words.

"A hotel."

"We are . . . waiting . . . for ladies, you sod." His head and torso wavered back and forth like a flag in a gentle breeze. "You . . . sod," he repeated with each sound drifting away.

They sat there staring at one another until Brian noticed Jeremy's eyes glaze over, then roll back into his head. He reached over to grab Jeremy's shoulder and guide his torso to the table as he passed out.

"What do we do now?" asked Jonathan.

Brian looked at his watch. Virginia and Anne would arrive any moment. "We can't take him anywhere. The girls will be here shortly. I guess we should wait for them, then we can take him to Virginia's or Anne's place."

"That would do, I should think."

"I've never seen anyone pass out like that."

"Me neither. He must have drank quite a bit of juice."

"I'd say."

Each retreated into his own thoughts considering what Jeremy had just told them. Alone, they wondered if the story was true. If it was true, what did it mean for the combat they were soon to face. Brian felt a chill shake his body. The consequences were staggering. Jonathan was the first to talk.

"Do you think it happened?"

"Jeremy was one of my flight instructors at Brize Norton. He protected me from some of the crap and helped me a great deal. He's never said anything untrue to me." Although he had withheld some of the truth, he told himself. "Plus, I've never seen him drink this much. I think he's telling the truth."

"My God, Brian. Do you know what that means?"

"I've been thinking the same thing."

"Here we are a few weeks into a war with Germany – the purported strongest military in the world – and we are shooting down our own aircraft, not Germans. My Lord above, what are we going to do when the Germans decide to really attack?"

Neither man wanted to answer the question. They drank their beer and stared at the blue uniformed hulk of Jeremy Morrison. The youthful vigor driving them to join an operational fighter squadron to fight Germans dampened somewhat as real life came to them both. In their own way, they both knew this was only the beginning.

The blanket of melancholy thrown over them by Jeremy pulled away slightly when Anne Booth and Virginia North walked into Shepherd's Pub. Brian received a delightful hug and kiss from Anne and even a kiss on the cheek from Virginia. Introductions were dispatched properly.

"What is wrong with Jeremy?" asked Virginia.

"He's had too much to drink."

"Why? He has never drunk so much he would pass out. Why?"

Brian knew he could not say a word about the Battle of Barking Creek, as Jeremy called it. The words did not come easily. "I think he's just had a bad time flying."

Virginia reacted quickly and strongly. "That has never bothered him before." A sense of suspicion rose within her. "Now, what has really happened?"

"Virginia, you'll have to ask him."

"From the looks of him, it will be tomorrow before I will be able to talk to him."

"Shouldn't we move him somewhere?" asked Anne.

"Your place is the closest," Virginia responded.

"Sure. Why don't we go to my flat? I have plenty of room."

"Maybe I should just get a hotel room," said Jonathan.

"Nonsense. As I said, I have plenty of room."

Jonathan looked to Brian who nodded his confirmation.

"Very well."

"Super. Now, let's get our friend out of here."

Brian and Jonathan got under each arm. Jeremy's body felt lifeless. Neither man realized how heavy a limp man could be.

The taxi ride to Anne's flat took less time than carrying Jeremy to the street and getting him into the taxi. The fog had thickened darkening the streets further.

"Let's get him to bed," Anne said as they entered.

"We should put him in the back bedroom," Virginia offered. "That way he will be closest to the commode. After all he has probably consumed, he is not likely to hold it down."

Lifting Jeremy up the narrow staircase proved the most difficult of tasks. In the end, Brian hefted Jeremy's body over his shoulder to finish the job and deposit him in the small bed. Jonathan waited in the front sitting room for the others to return.

"You have a simply delightful flat, Miss Booth," Jonathan said as he absorbed the refined décor including various forms of rather expensive artwork, numerous momentos, knicknacks, and photographs.

Brian took a more detailed assessment than he had at any previous occasion. The variety of well thought and personal items against the richness of the dark woods and exquisite wallpaper gave the room a deep elegance he could only describe as aristocratic although he had no direct means of measure. The sweet fragrance of lilacs amplified the dignity of the home.

"Thank you, Jonathan."

"Are you sure I am not a fifth wheel?"

"Not at all," answered Anne. "If you would like, I could ask one of my friends to come over."

The flash of surprise tainted with a strain of anger from Brian did not escape any of them. Brian's emotions were the most personal. He did not want Jonathan to know what Anne did to afford such an elegant town house. Jonathan seemed confused either by his expression or Anne's suggestion.

"No, thank you. But, maybe I should go."

Brian's emotions quickly changed from anger to embarrassment. "I'm sorry, Jonathan. Please, Anne means what she says. Please stay."

Jonathan looked carefully at each of them. "As you say, then."

"Good, now what do you say I make us a light supper."

"Not me," Virginia said. "I'm going to take care of Jeremy."

"I shall bring you up something in just a bit."

"That would be delicious. Thanks, Annie."

As Virginia disappeared, Anne went to the kitchen without a word. The two men stood in awkward silence neither one knowing how to talk about what was on their minds. The subject was the same. The perspective was completely different.

"What was bothering you back there, old bean?"

"Nothing."

"If looks could kill, I am afraid we would have a corpse on our hands."

"It was nothing, I tell you," Brian responded with more strength in his voice than he intended.

"As you say, then."

Hearing some of the interchange and sensing Brian's uneasiness, Anne called from the kitchen. "Brian, could you give me a hand, please?"

An intriguing blend of aromas greeted Brian as he entered the kitchen to find Anne with an apron covering her colorful dress and exquisite figure. She had nearly completed the tray of sandwiches, cheeses, pickles and other light edibles. It was not readily apparent what she needed help with. Only a moment later, Brian realized why Anne called him into the kitchen.

"I'm sorry, Anne, if I embarrassed you," he said softly.

Anne moved around the small table, stood close to Brian, rose up on her toes and whispered to Brian. "If you are ashamed of me, then I suggest a mutual parting to avoid further difficulties."

"Anne, please, I said I'm sorry. I don't know what struck me. I have always enjoyed my time with you." Brian hesitated wanting to say more. Anne waited. "I think I love you," he said lowering his head and eyes, and fidgeting with several utensils on the table.

"You do not have to say that."

"I am simply trying to tell you what I feel."

"Maybe so, but your expression back there was not one of love." Anne pulled back to make a last adjustment to the tray. "Now, let's eat," she said in a normal voice.

The tension of the first few minutes evaporated as they ate and talked of other things. The prospects of the war coming to England perforated an otherwise lighthearted conversation revolving mostly around the humor of the life adjustments brought on by the war. Anne did not believe anything would happen to England, nor would the war last long. Both young RAF pilots chose not to talk about the news and rumors they were aware of.

While Anne was reasonably familiar with the flying, she coaxed stories and experiences from both men. She seemed to have a genuine fascination with flight although she had never flown. Both men suggested they had to find a way to get her into the air, to enjoy the pleasures of flight and to see the countryside from the unique perspective of the birds.

The hour grew late. Fun, humor and laughter had a magical way of passing time. The awkward moment of transition was handled expertly by Anne Booth. Jonathan had his own bedroom toward the front of the first floor. Anne listened at the back bedroom door then knocked lightly.

Anne looked at the motionless body on the bed. In a barely audible voice, Virginia responded. "Everything is under control, Anne."

"Good night, then."

"Good night."

With Jonathan comfortably ensconced in his room, the two lovers retired to Anne's *boudoir* occupying the whole of the third floor. The exchange of feelings, emotions, concerns and sensitivities moved quickly through the broad spectrum of the afternoon's and evening's events. The words in time gave way to the passions of the flesh relegating any discomfort to the past.

Brian Drummond took great pleasure in demonstrating his feelings for Anne and his full appreciation of the things he had learned from her. In her own way, Anne could not resist continuing Brian's education on the fine art of lovemaking.

"May I be so bold to continue your apprenticeship regarding a woman's body?" Anne giggled in a friendly, playful way.

Brian's excitement over her suggestion took on a graphic form which Anne ignored. "Sure."

"Breasts are very important to a woman, Brian. The subtleties and sensitivities need to be appreciated and enhanced." She sat up in bed crossing her legs. "Here," she continued, taking his hand and placing it on her left breast. "Touch it ever so lightly. Feel the curves. Enjoy the texture. Feel how my nipple responds to your touch." They both watched as Brian continued his gentle exploration. The pinkish, flatness of her areola tightened becoming a darker, more reddish-brown, smaller and more textured like water logged skin. Her nipple rose up from the flesh of her breast becoming much larger and harder like a cylinder pushing out from the mound of her breast. "As things progress, you can squeeze and kneed it like bread dough. Take my nipple between your first two fingers and roll it. Harder. That's it. Ah, yes, that's it. A littler harder." She squirmed with obvious enjoyment. "Take my nipple in your mouth. Suck on it like a straw in a thick milkshake. Rhythmically. That's it. A little harder. Yes . . . ahhh . . . yes."

Brian did not miss her right hand reaching between her legs. A fascination enveloped him as her hand moved slowly at first. As he caressed and suckled at her breast, her hand took on a progressive urgency until she sucked in a deep breath holding it for the longest time, and then arched her back not quite enough to pull away from him. Her body shook quietly although he sensed the softest growl deep within her.

As the energy subsided, she fell back onto the bed straightening her legs in a moderate 'V'. When her eyes opened to look at him, she took his hand away from her breast and placed it between her legs. "Touch." She smiled. "Probe. Feel it?"

"I feel an incredible heat and slipperyness."

"Yes, but do you feel the contractions?"

"Yes."

"That is a woman's climax, Brian. That's what it feels like when you bring a woman pleasure."

"It's so fantastic."

"Yes, it is," she said relaxing letting him continue his exploration and education.

Anne Booth took advantage of a lull. "Brian," she said as she stroked the muscles of his chest and abdomen, "are you still upset at Jeremy for paying me to be with you the first time?"

"I'm getting over it."

"Maybe, you can take a turn that might make you feel better about it."

"Like what."

"Jeremy is unconscious, and Virginia is down there without much to do. What if I ask her to come up here with us?"

An image flashed into Brian's thought, but he shivered with the uncertainty of his inexperience. "You mean in bed?"

Anne smiled and ran her finger over his lips. "Yes, silly."

"I don't know. I've never done anything like that before. You do mean the three of us."

Anne nodded her head with patience and growing love. "Virginia is good. We shall have some fun."

"Jeremy is my friend."

"He won't mind. I should think he would be pleased. Remember, he has been with both Virginia and me for many years. Why wouldn't he want to share?"

"I wouldn't want to share you with them."

"You also have not lived the way Jeremy lives. It was just a suggestion for some spice of life, Brian. I don't particularly want to share you either, but

Virginia is my best friend."

The possibilities jolted his heart and his breathing. Anne could feel the reaction. She knew what the signs meant.

"I will go get her," Anne said as she rose.

Brian watched with simple fascination the subtle motions of her body as she walked across the room to the door. "You're not going down there stark naked, are you?"

"Jonathan is probably asleep. Anyway, I am certain I do not have any parts he has not seen before," she said with a broad smile illuminating her face, then disappeared out the door.

The possibilities were beyond Brian's imagination. The limitation did not stop the avalanche of thoughts. He did not have to wait long for the fog of inexperience to begin clearing.

The delicious unadorned lines of Anne's body led the taller Virginia wearing only Jeremy's uniform shirt. The gentle swing of Anne's breasts held his attention despite Virginia's presence.

"Are you ready for this?" asked Anne.

"I don't know."

"Relax This is going to be fun," Anne said as she pulled back the sheet covering Brian.

Her hands caressed him while she kissed him deeply. An edge of Brian's mind could not ignore the third person in the room. Anne stopped to look past the foot of the bed at her friend drawing Brian's eyes along with her.

With both sets of eyes on her, an impish grin accentuated the playfulness in Virginia's eyes as she slowly began to unbutton the shirt. With half the buttons undone, she pulled the top of the shirt away to expose her left breast. The two women took pleasure in the agonizing excitement they caused in Brian. Anne especially enjoyed the result gently stroking Brian to add to his physical sensations.

Virginia spread her arms and legs wide as if she reached some joyous finale. "What do you think, Brian?"

The feelings, thoughts, images and sensations of looking fully upon a tall, blond, blue eyed, well-endowed woman standing before him while being intimately caressed by an equally attractive woman overloaded Brian's capacity. The two women thoroughly appreciated Brian's release. The young man felt slightly embarrassed, but that passed quickly as the two friends continued to play. Hands, lips, legs and whole bodies moved gracefully about the bed. As is often the case with youth, Brian's state of readiness returned promptly under such stimulation.

"Now, lie back and enjoy," Anne said softly as she gently pushed Brian onto his back. "Virginia is the best there is at this."

Brian watched Anne's friend stand to full height on the bed. A serious expression of concentration washed over her that made Brian wonder what to expect next. She glanced briefly at Brian and Anne, then back to the object of her concentration.

"This is going to feel good," Virginia said more to herself than the others. She straddled Brian and lowered herself on to him expelling an audible groan as she took him fully inside her. "My, oh, my, aren't you a grand stallion."

The feelings of her movements and contractions were beyond Brian's comprehension as she took him to a new level of appreciation. Not wanting to miss out, Anne carefully and casually straddled Brian facing her friend. Brian did not need a cue as to what to do and used the skills he had learned from Anne to give pleasure to her as well as bring additional pleasure to himself.

Much to Brian's amazement, the two women kept after him through two full and violent releases before the positions changed. Through a wide spectrum of variations to their *ménage e' trois*, each of them enjoyed the satisfaction of several lofty peaks in sensual pleasure.

The fires of passion eventually took their natural subsidence. Brian knew he had seen more, felt more, smelled more and enjoyed more than he probably ever would. There was simply no way to have enjoyment and pleasure taken to such high levels. Even women of their experience appreciated the uniqueness of the evening's activities.

The sheets were damp from perspiration and the musk of human sexual pleasure as exhaustion subdued the three revelers. They lay entwined and motionless until the heat dissipated. Virginia rose gently trying not to disturb the other two which did not work. Both Brian and Anne watched Virginia retrieve Jeremy's shirt and leave the room dragging it behind her.

Much to her regret, Anne finally allowed herself to drop off into sleep with thoughts of Brian's reaction of the previous evening and whether he would be able to overcome his anxiety. She wanted him to succeed and hoped tonight's experience would help. She wanted her relationship with the American volunteer to continue. Her pure, unadulterated enjoyment of this man exceeded her plentiful experience. She did not want that enjoyment to end. Anne instinctively recognized there was much more to this man beside her. She was beginning to see some longevity to their closeness.

Tuesday, 26.September.1939
Bank of England
Threadneedle Street
City of London, London, England

BANK of England Governor Montagu Collet Norman, DSO, PC, waited impatiently for the scheduled private meeting with HMG Chancellor of the Exchequer. The declaration of war with Germany and the invasion of Poland added considerable complication to the bank's operations. He feigned disinterest with the pending meeting as he reviewed the endless data regarding the bank's operations. As his red pencil hovered above the page, Montagu's thoughts recalled the meeting with Chief Cashier K.O. Peppiatt six months ago, and he wondered if his underling had gone around him to the Chancellor. He believed not, as Peppiatt had always been a loyal colleague and it would not have taken this long for a political reaction. The war in Europe altered the banking landscapes despite his personal beliefs that it should not be so. Montagu knew that the banking system has to remain neutral and apolitical. He lived his life with that axiom of the banking business.

Missus Gathers knocked and entered. "Sir John Simon is here for his scheduled appointment with you, Governor Norman."

"Please show him in."

Missus Gathers held the door open wide. Norman rose from his large leather chair and walked around his desk to greet His Majesty's Government Chancellor of the Exchequer directly and personally. Chancellor Sir John Allsebrook Simon, GCSI, GCVO, OBE, PC, walked confidently into Norman's office and extended his right hand. The two men shook hands before the plush chairs arrayed around a short table before the office's unlit fireplace.

"Would you care for tea, Sir John?"

"Yes. That would be delightful. Thank you."

Norman looked to Gathers, nodded and waited for her acknowledgment. Simon chose the closest chair at one end of the table and sat. Norman went to the chair opposite Simon, at the other end of the table. The two economic and financial leaders traded cordialities until Missus Gathers returned with a silver tray, poured a cup for each man, prepared as each man requested, and closed the large office door behind her. Sir John took a bite of his biscuit and a sip of his tea, before he placed the cup, saucer and biscuit on the table. Norman waited for Simon.

"Please allow me to get directly to the point."

"By all means, Sir John."

"His Majesty's Government is nearly complete with our transition to a war footing. We need to ensure the central bank is properly aligned with the government." Simon paused and waited for an acknowledging nod from Norman. "To be specific, we are quite aware that the bank is the repository or agent of choice for numerous other international banks and nations. As such, we do not want the bank to allow or facilitate the transfer or liquidation of any assets, in any form, from or to Germany, Italy, Japan, or any occupied country involving the belligerent nations. My office will be notified of any communications with the qualifying countries, and any request for transfer or liquidation must be approved by the Chancellor or the Prime Minister. Do you have any questions?"

"Yes, I do, as a matter of fact. This is an extraordinary . . . I do believe unprecedented . . . action on the part of His Majesty's Government. Under what authority does the government take this action?"

"The Emergency Powers Act passed last month."

Governor Norman stared at Chancellor Simon as he organized the words he knew he must speak at this pivotal moment. "Sir John, you know my approach to banking policy. We have discussed it many times. I know you agree in principle with the notion of banking neutrality and remaining apolitical. I can only imagine the enormous pressure that must have been exerted upon you to overwhelm your general, broad agreement. As such, I urge you to reconsider this draconian edict. This is not the way to run a major bank."

"I expect nothing less in your position, Montagu. However, I must remind you, these are not normal times and the war we have just entered will likely not be your average armed conflict, even when viewed by the standard of horrific carnage and destruction of two decades ago."

"Yes, but . . ." Norman stopped when Simon gave him the universal sign.

"His Majesty's Government cannot tolerate any assistance to the enemy. Certainly you see that currency resources and other assets are or can be a powerful, enabling instrument."

"Yes, I do appreciate the dilemma; however, by succumbing to this directive, we will compromise the integrity and moral standard of this grand institution."

Sir John lifted his cup and took several sips, as he considered his choice of words. "The current German regime has demonstrated repeatedly their paucity of any moral grounding. I shall not recount the evidence for your education, Montagu. His Majesty's Government expects you to perform your duties as a loyal servant of the Crown."

Norman felt anger jacking up his blood pressure and undoubtedly flushing his cheeks not covered by whiskers. "If we had not been long term colleagues, I would surely have taken offense at your pejorative tone, Sir John. My position is not personal; it is my professional counsel to His Majesty and the government that serves his Crown. I shall not defend my loyalty, as I feel no need."

"No need to get all twisted up, Montagu. No one is questioning your loyalty. We simply do not want our financial institutions, directly or indirectly, providing aid to the enemy. Surely, you can appreciate this point."

"I think this is a mistake that potentially has profound ramifications. Is His Majesty's Government willing to sacrifice the future for a modicum of leverage in the present?"

"If there is no future, the question is moot."

"You can't be serious."

Sir John stood and walked across the room to look out the window on the cityscape around them and hustle of humanity beneath them. After several minutes, he walked back and stood behind his chair, facing Norman. "I did not expect this to be easy, as I am well aware of your position. In general, I agree with you. The banking industry should not take political sides. Nonetheless, the Emergency Powers Act gives His Majesty's Government the authority to take all necessary steps to protect the Home Islands and the British Empire. The Prime Minister and the War Cabinet, including the Chancellor of the Exchequer, believe this is a necessary and essential step toward that objective."

Norman recognized the rope was becoming taut and he did not have much more space. "Are you aware of the BIS order we executed early last spring?"

"I am afraid not," Sir John answered, and then sat down.

"We received a proper request from BIS to transfer £5.6 million worth of gold from one of their accounts to another."

"That sounds rather routine for the Bank of England."

"Yes, well, our chief cashier brought the request to my attention and suspected the transfer was from the National Bank of Czechoslovakia to the German Central Bank."

"Last spring, you say?"

"Yes, one week after German troops marched into Prague."

Sir John Simon stared at Montagu Norman like a stone statue. "What did you do?"

"It was a proper order between two BIS accounts in our custody, and I had no other instructions. We executed the order. Further, the owner of the receiver account sold the gold in just over a week after the transfer."

"You did not consider that perhaps you should have consulted with me before hand?"

"I saw no need to bother you with a routine transfer request from one BIS account to another."

"That was looted gold, Montagu."

Norman studied Sir John's face, which offered no clues. "With no intention to be presumptuous, how would you know that?"

"By your words."

"Sir John, I used the word suspected. I did not use evidence with certainty or any variation thereof. We are operating within the ambiguity of the anonymous international banking business. We do not have access to the client account registry of the Bank of International Settlements, or any other foreign bank. To deal with situations comparable to ours, we need a body of law among nations."

"You make a valid point. Nonetheless, we do not have those agreements or tools in place, and the directive of His Majesty's Government is clear, unequivocal and not negotiable. I trust you will respect the directive."

"I suspected this was the direction we were headed. I received a nearly identical request just last Friday and asked our chief cashier to delay execution until we had this meeting."

"Same two accounts? How much?"

"Yes, exactly. This request was for the transfer of about forty thousand kilos of gold."

"I shall inform the Prime Minister and War Cabinet. Until I indicate otherwise, please leave this request in a pending status. Let us see how long BIS-Basel will let it go."

"As you wish, Sir John."

"Thank you ever so much."

"Should we notify our clients?"

"I would say not. We would rather deal with each request both as a potential bank transaction but also as a source of intelligence. If we truncate activity at the source, we lose both."

"Yes, quite right. Then, so be it. The bank shall comply and cooperate with His Majesty's Government."

"Excellent. Thank you, Montagu. If you would be so kind, please convey the government's gratitude to your chief cashier for raising his suspicions last spring."

"I will be honored to so inform Kenneth Peppiatt. He is a good man and holder of the Military Cross from the war in France."

"Please thank Mister Peppiatt for his service to the Crown on and off the battlefield." He nodded. "Thank you, again, Montagu," Sir John said as he stood and extended his right hand.

The two men shook hands and exchanged their friendly farewells. As the door closed, Norman stared at the door as his thoughts raced through the potential consequences of what had just happened. Several minutes passed before he walked to the window to observe the departure of the Chancellor of the Exchequer. As his automobile disappeared down Threadneedle Street, Governor Montagu Norman returned to his desk and the tasks awaiting his review or action.

—

Friday, 29.September.1939
The Admiralty
Whitehall, London, England

THREE men, leaders of the Royal Navy, sat around the office conference table of the First Lord of the Admiralty. The First Sea Lord and the Director of Naval Intelligence joined their principal minister for the hastily called midday meeting. Each of the professional naval officers had been asked to gather the latest information they possess regarding the current situation. Winston just returned from a morning War Cabinet meeting at No.10 Downing Street. The news had not been good.

"Well, gentlemen," he said after the usual amenities, "we have a different Europe. As you well know, the Russians invaded Eastern Poland twelve days ago across virtually an undefended frontier. It appears they were quite successful in their stab at the back of the Polish Army turned to face the German onslaught. At any rate, the Foreign Minister reported to the War Cabinet this morning that Russia and Germany signed a treaty formally partitioning Poland. They have carved up the carcass. Do we know any more?"

The intelligence question fell to Sir Geoffrey. "Pockets of resistance, especially in Central and Southern Poland, continue to cause problems for the Germans. However, according to SIS, the *Wehrmacht* has done its part. They have deployed SS units that are progressing rapidly to consolidate their spoils."

"Has there been any change in German naval deployments?"

"None we can detect. Completion activity on the *Bismarck* continues to progress unfortunately quite well. The *Graf Spee* sunk her fourth merchantman two days ago."

"We have three battle groups hunting her," interjected Admiral Pound.

"We might expect the *Deutschland* to attempt a break out into the Atlantic any day now."

Winston looked to Admiral Pound. "Do we have sufficient forces deployed to seal off the North Sea?"

"No, sir. As you know, we are making needed repairs to several capital ships. We have an insufficient number on the line to assure interdiction should they attempt a break out. The Channel is effectively sealed although the Channel Squadron is not heavily gunned for a free surface battle. Given the confines of the Channel, they could handle either battleship. It is the Northern approaches that are the most vulnerable."

"Does the cessation of hostilities in Poland and the treaty with Russia change our battle plans?"

"I should not think so," answered Admiral Pound. "While the Germans should be able to free up a flotilla or two of destroyers and a few submarines, we can see no evidence that might lead us to alter our deployments or battle plans."

"Is there any other relevant information I should have before I walk down to Commons?" Both officers searched their memory and eventually shook their heads to signal a negative. "Very well, then, we shall redouble our efforts to anticipate any German moves across the Baltic or into the North Sea." Winston asked himself if there were any other topics that needed to be addressed. "Ah, yes, one additional question that has been plaguing me lately and I forgotten to ask. Do you recall that unfortunate incident over Mersea Island?" he asked referring to the RAF's Battle of Barking Creek. Both men nodded their heads. "Do we stand the same risk?"

Admiral Pound answered. "Not likely, although I suppose in the heat of battle, the failure to recognize a friendly vessel is always a possibility."

"Do we practice enemy vessel and aircraft recognition?"

"Yes, certainly, as does the RAF."

"Their recognition skills were not sufficient in that regrettable episode."

"As Sir Cyril noted, First Lord, there were many contributing factors. The lads got excited in a confused situation and were a smidgen too quick to react. While the same is certainly possible at sea, it is less likely. We do not have single seat battleships."

They laughed at the image. The response satisfied the First Lord of the Admiralty. He looked over his shoulder to the large wall clock next to his familiar world-wide deployment map for the Royal Navy.

"I should be off. We stand adjourned. I suspect we shall have a late night in Commons. No need to wait for me."

"Thank you, sir," the two admirals responded simultaneously.

They left Winston to gather up the few papers he felt he needed for the afternoon Commons session. In short order, the First Lord left his office and the Admiralty for his usual unescorted walk to Westminster Palace. The growing shadows offered some added coolness to the early autumn day. He noticed his friend, Anthony Eden, emerge from the War Office on the south side of the Whitehall thoroughfare. He judged traffic in both directions, then crossed the street.

"Good afternoon, Anthony," he called to his friend several yards ahead of him. Eden turned to find the caller, and spotted Winston moving quickly toward him. "How has your day been, if I may ask?" They walked slowly down Whitehall.

"Reasonably good, despite the news, thank you. And, yours?"

"Reasonably good as well, I should say."

"What do you think about this grotesque relationship between Soviet Russia and Nazi Germany? They are such an odd couple."

"They certainly are . . . from opposite extremes of the political spectrum. Hitler has used the hatred of communists to mask his rabid attitudes toward the Jews. I can find no solid justification for this relationship other than perhaps time."

They exchanged greetings with citizens who passed them on the street. Both men enjoyed being among the populace especially when the greetings were positive, cordial and appreciative.

"How long do you think this will last?"

"As long as it takes to subdue France and Britain. Assuming Hitler is able to lay claim to Western Europe, he will then shift the objective of his voracious appetite and megalomaniacal ambitions to the East. Their turn will come, of that I am certain."

"Between us and no one else, will the Germans really take on the French, much less defeat them in battle?"

"Not only to I believe he will attack France within a year, he will most probably be victorious unless by some miracle he gives us time to prepare. We need two to three years if we can catch them at all. Although I will state so publicly or to anyone other than my closest friends, the French Army is a facade. They have the numbers of men, but their equipment is vastly inferior to the German's, as ours is as well. As far as I can tell, German armor is unmatched by any in the world. The Americans are even further behind than we are. If Hitler decides to unleash his armor generals, we shall bear witness to the most innovative, violent and aggressive warfare wrought by man upon his kind."

"Why upon God's Earth have the Russians joined these Fascists?"

"From my perspective, Soviet Russia is a riddle wrapped in a mystery inside an enigma. The only perceptive key is Russian national interests. Somehow Stalin must believe this action buys him time which they need as well. They have part of Poland. I surmise they would like to have a buffer between them and Germany they can obtain by any means."

"That would make sense."

"It is the only thing I can figure."

"The War Office seems to be quite confident they will be able to handle their part."

"False belief. I am afraid they do not realize what they shall soon face."

"Then, how do we defend Mother England?"

"The way we always have . . . with the Royal Navy placed squarely in our moat."

"So, we should consider ourselves beyond reach?"

"*Au contraire, mon ami*. The difference this time is aviation. As we saw with the Zeppelins in the last war and certainly watching the *Luftwaffe* in Spain and Poland, this time aviation is far more mature. To an even greater extent, our fate shall rest in the youthful hands of our pink cheeked fighter pilots. They must maintain supremacy in the air or our moat will surely be bridged for the first time in many centuries."

"The Army does not see it that way," observed Eden as they approached Bridge Street, Parliament Square and the glorious detail of Westminster Palace.

There were always people around Parliament Square. The attraction of Westminster Palace as well as Westminster Cathedral and the various buildings of government provided the magnetism for people virtually year around. They returned words of encouragement and praise as they made their way across Bridge Street and onto the Palace grounds.

As the public words faded behind them, Churchill continued their conversation. "Unfortunately, there are not enough visionaries in the Army. Some of the generals learned the wrong lessons in France and Flanders fields. We have done the best we can do. My only prayer is that we are able to weather the first cracks of the storm. I am truly confident we shall prevail, if we can survive the first few months."

"We must find a way, then, mustn't we?"

"Precisely."

"The business of government awaits us, Winston. Let us place our shoulder to the wheel," said Eden with a cheery, up tone to his words.

"Excellent. As you say, then," Winston responded as they passed through the Member's entrance into the historic hall.

Chapter 13

> Each friend represents a world in us,
> a world possibly not born until they arrive,
> and it is only by this meeting that a new world is born.
>
> -- Anaïs Nin

Friday, 20.October.1939
RAF Hawarden
Broughton, Clwyd, Wales

"Well, Brian, today is the day. It should not be, but it is," announced Pilot Candidate Derek Langston, but the activities remained the same.

"We should tick over a little faster if we are going to get anything to eat," Pilot Candidate James Roland suggested.

The pilot candidates of D Flight moved quickly like machines. Words of camaraderie kept humor in their lives no matter what the weather, obstacles or tasks in front of them. The jokes and gibes among friends always made the cold warmer and the rain softer.

After breakfast as they prepared for their flights or other duties, Ian Milner returned from his stint as the assistant to the squadron duty officer, more commonly called, the gopher – go-for-this, go-for-that. Since the declaration of war, the young pilots gained some modicum of satisfaction from the night duty. It was the only time there was enough time to keep up with events of the war. For the fledgling warriors, the events of war held a certain irresistible attraction. As each one of them finished a night tour, the remainder would interrogate the wiry pilot candidate.

"Here is our mate. So, what happened yesterday, Ian."

"Can't we do this later. I am absolutely exhausted."

"No, we can't. You can sleep whilst the rest of us are flying or something."

"Maybe a little then. I'll fill in the details after the graduation ceremony this afternoon."

"Deal."

"Flight Lieutenant Johnstone thinks the Germans will attack in the North."

"Why is that?"

"The Germans flew numerous reconnaissance flights over the North country yesterday and last night. We also received confirmation a Heinkel One One One was shot down by Spitfires over Rosyth."

"Where is that?" asked Derek.

"You have led such a sheltered life," quipped Peter. "Rosyth is in Scotland, near Edinburgh."

"So, how does Johnstone think one Heinkel means the Nazis," Derek said in a good imitation of Winston Churchill's intonation and pronunciation, "are going to invade the North?"

"HMS *Royal Oak* was sunk by a sodding German submarine at bloody anchor at Scapa Flow six days ago. The daily recce flights. Jesus wept, they have even dropped sodding bombs. What have they done in the South? Nothing."

"For all you know."

"Agreed. We are just candidates and we probably do not know, but Johnstone and the other instructors are fighter pilots."

"Most, not all," interjected James.

"All right, so stretch me out for a little exaggeration."

"What else?" asked Brian eager to get back to the details.

"It is the logical place. Less people, lighter defenses."

"And more rugged terrain."

"My, my, aren't we a critical lot of odds and sods? That is enough. You blokes get out of here so I can get some sleep. I will catch up at the ceremony for our darling Brian."

For D Flight, this was an ordinary day filled with classroom lessons on techniques and procedures and with training flights across a broad spectrum of flying skills. For Brian, as a graduating candidate, he would fly one last flight in OTU7 – a flight with the unit commander, a tradition within the organization. Brian and three other pilots would be going through the same set of activities. Once the flights were completed, the ceremony in the afternoon celebrated their promotion to the rank of pilot officer – the equivalent of second lieutenant in the Army or ensign in the Navy – as they received the winged insignia of a pilot in the RAF.

When Brian reported to OTU7, the ceremonies took place once every other week or so. Now, they occurred every Friday as more pilots completed the course. The uneasiness of anticipation placed enormous pressure on the RAF Training Command to produce pilots. There were more seats in Fighter Command and Bomber Command than there were pilots to fill those seats. The pressure would grow far worse before any sign of relief could be felt.

The traditional flight with Wing Commander Martin Norstreet had become a signature event used as a capstone to a difficult, intense six month advanced flight training program. The flight touched each of the primary skill areas, basic airmanship, navigation, formation, gunnery, and aerial combat. Norstreet enjoyed the final flight routine. It was his excuse for flight time. He also took great pleasure observing the graphic demonstration of the capabilities of the next generation of RAF pilots. Wing Commander Norstreet looked forward with special anticipation to one of the four final flights on this Friday.

Flying Officer George Esling, one of the better instructors in OTU7, reported exceptional scores in most skill areas. Everyone knew Pilot Candidate Brian Drummond scored the highest recorded hits on the banner targets. George spoke very highly of Drummond although he kept his opinions within the closed professional circle of instructor pilots. Norstreet wanted to see for himself.

"Mister Drummond, are you ready for your final evaluation flight?" asked the commander.

"Yes, sir." The final flight was a good time. The hard part was over.

"Excellent. Then, I suggest you fetch your kit and let us be off."

Recognition that this flight was different came quickly to Brian. As the Gladiator's wheels broke the ground, Norstreet radioed from his aircraft just behind the left wing. "Your oil pressure has dropped to zero, and the engine has begun to run rough."

The routine was quite familiar to Brian although it was certainly unexpected. "Hawarden Tower, Sword Two One. I have a practice engine problem, and I'm returning to land." Brian kept the throttle up knowing that was his best chance if the engine was truly failing. He made an immediate right, climbing turn until he had the nose pointed back at the large grass landing area. He set up for a crosswind landing that he executed a perfect landing in the light crosswind.

Norstreet was impressed with the precision of Brian's flying. The remainder of the flight progressed at the same level as the first simulated emergency. Brian Drummond dazzled his instructors with his gunnery skills, and he continued to do the same with Wing Commander Norstreet. As the flight proceeded, the senior RAF pilot became an advocate. He knew the young American was good and that His Majesty's armed forces needed many more pilots like him. Even Brian's difficulties with formation flying early in the program had disappeared and now become a superior demonstrable skill. While he knew there were particular elements he could do better, Brian felt good about his performance and his ability to serve as an officer in the Royal Air Force.

With the aircraft completely shutdown, Brian quickly extricated himself from the Gladiator and moved swiftly to wait for Wing Commander Norstreet to join him on the ground. No words were exchanged. Not even a smile passed between them after Norstreet told the crew chief about several aircraft problems, called squawks, and he began walking back to the Operations building. Brian walked the customary half step behind and to the left of the senior officer. Norstreet closed out the flight and walked to his office. Pilot Candidate Drummond guessed he was supposed to follow. Once inside, Norstreet closed the door.

"Take a seat, Mister Drummond."

Brian complied.

"I will tell you this in private and I will deny it if you mention my words to a bloody soul."

Brian nodded his acceptance.

"You are one of those rare instinctive pilots who has a brain and a hunter's mind. While it is well known I am not particularly fond of foreigners, I am not sufficiently stupid to deny this country's need for fighter pilots of your caliber. You are among a unique class of pilots and you should be proud. Although I usually refrain from providing loose advice, in this case, I thought I would offer one piece of advice. You are a guest in this country, a very important guest, but a guest nonetheless. You will find your service with the RAF and your stay in Great Britain far more enjoyable if you keep your head down. Do you understand?"

"Yes, sir."

"Brilliant. Then, we should have no problems whatsoever, now will we."

"No sir."

"Two questions, if I might." Brian nodded. "Where did you learn to fly?"

"With a friend in my home town."

"You have learned well and your performance certainly speaks well of his instruction."

"One last question to satisfy my curiosity. Who is your benefactor within the RAF?"

The question confused Brian Drummond. Did he want to know who got him into the RAF? Who his instructors were? Could he be referring to Jeremy Morrison? "What do you mean, sir?"

"You must know somebody important?"

"Flight Lieutenant Lord Jeremy Morrison?"

Wing Commander Norstreet's laugh confused Brian even more. Had he said something wrong, offensive or ridiculous? "No. Somebody much higher?"

"Group Captain Spencer?"

"Now, we may be getting there. I presume you are referring to John

Spencer, Churchill's nephew, and Dowding's Staff Secretary?"

"Yes, sir."

"Then, I think we have a winner. Your posting from the Training Command is to Six Oh Nine Squadron." Brian's expression told Norstreet the American did not understand. "Six Oh Nine is a volunteer reserve fighter squadron based at Drem, Scotland, near Edinburgh, and they fly Spitfires."

"Yee-haw," shouted Brian in the uniquely American Western manner.

"Mister Drummond, please, a modicum of control and reservation would be proper and greatly appreciated."

"Yes, sir. Sorry, sir," he responded with obvious excitement.

"I can count on one hand the number of instances of men training in one aircraft and being assigned to another. Intervention seems to be the only means."

"Yes, sir." Brian still struggled trying to contain his excitement. The prospect of flying Spitfires as opposed to Gladiators made his task of control immensely more difficult. All his thoughts drew inward to the sleek, graceful lines of the aircraft he first saw four months ago. The urge to call Group Captain Spencer to thank him was very strong.

"Well, then, I am glad you are pleased." Norstreet stood to come around the desk and extend his hand to Brian. "Congratulations and thank you for a most enjoyable evaluation flight. Fly the damn thing to its limits and kill lots of Germans."

"Yes, sir."

"Now, get the hell out of my office, you bloody Yank." Norstreet smiled a friendly sort of smile. "I shall see you again at the ceremony."

"Yes, sir," Brian responded crisply, turned on his heels and departed the Commander's office and the Operations building with the broadest of smiles on his face.

The two hours until the graduation ceremony provided no outlet for Brian's excitement. All his mates in D Flight were performing the requisite tasks in training. The first to return to the residence building was James Roland. Brian simply could not contain his excitement as he told each of his mates as they returned to the building. By the time they needed to leave for the assembly hall, everyone in D Flight knew Brian would soon travel to Scotland to join a Spitfire squadron. They were excited, envious and jealous all at the same time. Each of them handled the emotions differently. Every young RAF pilot aspiring to fly single seat fighters wanted the new, sexy, sweet sounding Spitfire. Most would accept Hurricanes. None wanted to fly Gladiators, Blenheims, Defiants or relatively old Furys.

The ceremony became a most memorable event for Brian Drummond. For only the second time in his young life, he was publicly recognized for his flying skills. This time, more than winning fourth place at the Oklahoma City Airmeet, in addition to finishing advanced flight training with one of the highest scores, Brian received the wings of a RAF pilot and a full commission as a pilot officer. At this moment, becoming an officer had more meaning to him. Numerous people of various ranks made sure he knew his receiving a commission without a college degree was most unusual. Wing Commander Norstreet told him it was his connections. Flying Officer George Esling, his principal flight instructor, told him it was in recognition for the exceptional flying skills and obvious time spent in prior training. Whatever the reason, Brian Drummond was quite thankful.

"Let's get pissed," shouted Peter Baker as they approached the Officer's Mess.

"Easy as you go, Peter," said Derek Langston. "Some of us have to fly tomorrow."

"Quite right," added Jonathan Kensington.

"Ah, you short strokers are all alike. No stamina."

"We just want to stay alive."

Peter grunted and grumbled as they entered the bar. The room was crowded as usual and filled with the noise of several scores of conversations. A pint of bitters was thrust into their hands by several candidates from E Flight. Congratulations in various forms occupied the early portion of the post-ceremony celebration.

"What are you going to do with your week's holiday?" asked Ian Milner.

Brian smiled broadly. "I'm going straight to Drem to join my squadron."

"My God, man, are you absolutely bonkers?"

"Don't be stupid."

Numerous other personal observations and comments joined the choir. New resolve merged with his commitment. A week's holiday could not compete with his first flight in a Spitfire. He had waited a long time to reach this point and time off simply was not important.

"Are you going to take some time with Anne?" asked Jonathan with a subdued voice so the others could not hear.

As it happened every time previously, Anne's name conjured up vivid images of her elegant face, exquisite body and delightful feel. This time those images were confused by the deep purr of a Merlin engine at full throttle and the graceful curves of the fuselage and wing of the Vickers Supermarine Spitfire Mark I single seat fighter airplane. "No," responded Brian, "not this time, as much as I would like to."

"Brian, it cannot be that important."

"It is to me. It's all I've dreamed about for more than a year."

"To pass up a yummy go at that gorgeous woman?"

"I'll see her my next break."

"Do not forget, Yank, Drem is twice as far from London as Broughton."

"I hadn't thought of that, but it doesn't change my mind."

"When are you going to leave?"

"Tomorrow morning."

"Furthermore, the North appears to be where most of the action is."

"I hadn't thought of that, either. Even more reason for me to go straight away."

"It's daft, Brian, but it is your life and pleasure."

Since everyone except Brian had to fly Saturday, no one including the vociferous Peter Baker could muster up the energy to go into Liverpool as they had done so many times before. As had become the norm since the war began, the range of their party spirits decreased substantially although by no means disappeared.

The smoke in the bar combined with the beer to dampen Brian's enthusiasm for the party. He knew Jonathan felt the same way. They both began to work their way toward the door.

"Wait up there, Mister Drummond," said Flying Officer Esling. "Could I have a word with you?" he asked meaning alone.

"Yes, sir."

"I will see you back at the shed," offered Jonathan.

They walked outside and allowed Jonathan to separate by three meters. "I may never see you again," began George Esling. "I am not sure who was instructing whom sometimes with you, Drummond. I learned a heap from you. You are one hell of a pilot. You will fly the bollocks off the Germans is my guess."

"Thank you, sir. I learned a great deal from you. I appreciate your patience and extra work with me especially with improving my formation skills."

"Good luck, Brian."

A light drizzle added dampness and a chill to the autumn air. The crackle of fallen leaves beneath his feet accentuated the rhythm of his pace. He wanted the night to pass quickly. The journey to Drem and his future could come none too soon.

Friday, 20.October.1939
The White House
Washington, District of Columbia, USA

"Mister President, we just received this Eyes Only cable," announced Missy LeHand, Franklin Roosevelt's executive secretary.

"Thank you, Missy."

Roosevelt extracted the single piece of yellow paper with its distinctive teletype printed letters from the envelope. It was from Winston Churchill which meant this was the first communication under the agreement the two men reached through William Stephenson, otherwise known as 'Intrepid.'

MOST SECRET

```
1
20.X.39
TO:    POTUS
       AS THIS IS OUR FIRST MESSAGE, PLEASE
ALLOW ME TO TAKE THIS OPPORTUNITY TO THANK
YOU FOR GENEROUSLY CONSENTING TO THIS RATHER
UNUSUAL CHANNEL OF COMMUNICATION.  AS I
INDICATED THROUGH INTREPID, THESE TIMES DEMAND
UNIQUE ENDEAVOURS.
       WHILE I KNOW YOU ARE AWARE OF NAZI
ACTIVITIES ON LAND AND AT SEA, I WANT YOU TO
KNOW HMG IN CONJUNCTION WITH THE REPUBLIC OF
FRANCE ARE MAKING EVERY EFFORT TO AVOID ALL OUT
WAR WITH THE GERMANS IN HOPES HITLER WILL COME
TO HIS SENSES AND WITHDRAW FROM POLAND.  THIS
IS NOT A LIKELY OUTCOME.
       THE ROYAL NAVY WILL CONTINUE TO PROSECUTE
ANY ENEMY TARGET OR THREAT.  EVERY EFFORT WILL
BE MADE TO RENDER ASSISTANCE TO NEUTRAL OR NON-
BELLIGERENT SHIPPING.  HOWEVER, AS I AM CERTAIN
YOU RECOGNIZE, THE NAVY CANNOT BE EVERYWHERE
AND I SUSPECT THE NAZIS WILL IN TIME ATTACK
EVEN AMERICAN SHIPPING.
        I SURMISE HITLER WILL USE THE WINTER TO
CONSOLIDATE HIS SPOILS IN POLAND AND PREPARE
FOR A FURTHER EASTWARD PUSH IN THE SPRING.  AN
```

> UNPROVOKED ATTACK ON THE LOW COUNTRIES AND
> FRANCE IS ALSO A DISTINCT POSSIBILITY. IF
> EITHER OF THESE EVENTS SHOULD OCCUR, THE
> WHOLE OF EUROPE WILL SOON BE FULLY ENGAGED IN
> WAR. FURTHERMORE, INSULATION OF AMERICA FROM
> THE PERVERSITY OF MODERN WARFARE SEEMS QUITE
> UNLIKELY.
>
>   SIGNED:
>   NAVAL PERSON
>
> **MOST SECRET**

The President of the United States reached for the intercom box on his large mahogany desk. Pressing the appropriate lever, he said, "Missy, please find Harry Hopkins. I need to talk to him."

"Yes, sir."

Roosevelt reread the message and considered the information within it. The domestic pressure to remain neutral in this war continued at a high level. The vocal isolationists made any relationship with the British politically risky even for a popular president. The thought of American shipping being sunk by German submarines brought even less comfort. While Churchill's opinions were cogent and logical, he did not share the pessimistic view. His opinion along with most of his advisors held that Hitler would probably not risk provoking the United States. Churchill's views about the importance of the United States in the world arena were clearly stated, public and well understood. He had to take any message with a grain of salt no matter how much respect he deserved.

The buzzer on the intercom box drew his attention to the small red light over the third lever from the left on the top row. "Yes, Missy."

"Mister Hopkins is at the State Department with the Secretary of State. He will be here in twenty minutes."

"Thank you."

Roosevelt extracted a small pad of paper. With pencil in hand, the President thought carefully about his response to the First Lord of the Admiralty.

Missy LeHand typed the message from the President's handwritten note and returned it to him.

> **TOP SECRET**
>
> 2
>   October 20, 1939
> To: Naval Person
>   Thank you for your most informative
> message. Concern over the intentions of the
> German Chancellor remains acute. The thought
> of American citizens engaged in this conflict
> is not particularly attractive. Likewise, the
> possibility of a Nazi empire encompassing all
> of Europe is not acceptable.
>   It is my intention to provide whatever
> assistance I am able to offer within the laws
> set forth by Congress. Sacrificing my public
> position for any cause, however noble, would
> not be productive or contributory to the cause.
>   Let us continue to correspond and find
> ways we may be of mutual assistance.
> Sincerely yours:
> Former Naval Person
>
> **TOP SECRET**

President Franklin D. Roosevelt reread his note twice with a review of Churchill's message in between the two readings. He was satisfied his response provided encouragement without specific commitment. Informal and unofficial communications of this type no matter how noble of purpose could be immensely damaging in the wrong hands. The strength and sustainability of the British Isles in the face of the German war machine had to be considered questionable. Roosevelt found no more comfort than Churchill that the distinguished Member of Parliament had been precisely correct about Hitler and the need for military strength in the face of the threat.

As his mind moved to the text of a pending farm subsidy bill before the House of Representatives, the buzzer rang again.

"Yes, Missy."

"Mister Hopkins is here."

"Show him in, please."

The President retrieved his note to Churchill from the right hand drawer. "Good afternoon, Harry," Roosevelt greeted his friend, confidante and advisor.

"Good afternoon to you, Mister President."

"Harry, come on over here. I've got a couple of pieces of paper for you to read." The President handed Hopkins first Churchill's message, then his penned response. "What do you think?"

"I'd say our friend Winston is a rather clever and ballsy man."

The shared laugh complemented the words. They both felt the same.

"Indeed. You have to admire the man. It is as if he intends to win this fight despite the government."

"You're right, Franklin."

"If you were Winston and received this response," he said holding up the piece of paper, "how would you interpret my note?"

Harry Hopkins contemplated the question and his answer. "I'd say you were sympathetic but not enthusiastic about support."

"I'm afraid that is the best I can do at this juncture."

"Yes, sir."

Roosevelt took one last look at the return message. "Would you get this to the Communications Center?"

"Right away."

"Thank you for quick response, Harry."

"Always a pleasure."

―

Saturday, 21.October.1939
Edinburgh, Lothian, Scotland

The chief conductor's announcement approaching Edinburgh's Waverly Station brought a touch of excitement. Brian knew he was getting closer to his squadron and his first flight in the beautiful fighter. The rain and wind made the exterior seem so cold while the interior of the rail coach was warm. He traveled virtually alone on the rail journey from Chester.

The seas of green with a healthy mixture of autumn colors amazed and impressed Pilot Officer Brian Drummond. The English and Scottish countrysides were the most beautiful he had ever seen. The hours of scenery passing by the window of his rail coach compartment on this Saturday had stimulated his reminiscence. The Kansas countryside would be shades of brown at this time of the year. Inevitably, with quiet time to consider, the fading vision of Rebecca Seward brought strings of guilt and some remorse that could not stand against the intimate power and presence of Anne Booth. His thoughts also returned to the most recent letters from Malcolm Bainbridge and his parents.

Malcolm's last letter confirmed what his parents had told him in their letters. Susan and George Drummond had already contacted the United States State Department through their senators and representative. They intended to remove Brian from the RAF and return him to America using the Neutrality Act. Malcolm said he could not get involved in a family affair, but Brian should be prepared for whatever might be ahead. The thought of his parents trying to get him removed from the RAF caused him to feel some resentment. He wanted no interference that might affect his dream. His parents seemed to ignore his feelings, desires and the essence of his dream to fly with the best. His frustration with not being able to get through to them added to the burden he carried. The meddling efforts of his parents were one main reason he was so willing to forgo the proffered week's holiday to jump into the Spitfire as soon as he possibly could. Brian wanted to tell Jonathan but decided to hold back. If he was going to be forced to return to the U.S., he wanted to at least have flown the fastest, most maneuverable, best looking fighter airplane in the world.

A jolt of the rail carriage changing tracks drew his attention out the window. Instantly, Brian noticed the city rising on either side of the rail line. As he leaned down to see the top of a rock face out the right side of the coach, the impressive structure of Edinburgh Castle and its walls built on top of the sheer rock face struck Brian with awe. The regal castle perched at the pinnacle of the precipice in the center of the old city dominated the scene. Brian knew without question this city belonged in fairy tales and probably was in some story or another. He was thankful his flying passion provided a steady progression of incredible experiences from the cities of the Great Plains, to the university city of Oxford, to the famous port city of Liverpool, and now the majesty of Edinburgh Castle.

"First time to the queen city of Scotland, laddie?" asked an elderly and craggy man with a heavy Scottish brogue that was difficult to understand.

"Yes."

"You're one of those flyboys, aren't ya?"

"Yes, I am."

"Where ya from?"

"America."

"Well, I'll be. A colonial. Come to save ol'Britain, have ya now?"

"I suppose. Really, I just want to fly."

"Then, ya must be goin' down to the village Drem, then."

"Right, again. Maybe you could tell me the best way to get to RAF Drem?"

"I certainly can, laddie. The next London Midland and Scottish Railway train to Newcastle," the old man paused to check the schedule board, "leaves at just after one this afternoon," then he checked the clock at the far side of the station, "twenty minutes from now."

"Thank you, sir."

"Thank you, laddie. We need strapping young lads such as yourself to whip these bloody Gerries."

"We'll do our best."

"I know ya will," he added as he turned to depart. "By the way, should be the third stop. Drem is the station. Drem, that is, now, laddie."

"Yes, sir. Thank you."

Brian walked to the edge of the station house and took a deep breath of the damp, cool air heavy with pungent odors of oil, soot and steam. He gazed up at the impressive castle and the equally impressive stone buildings of the old city. "My God, what a place," he muttered aloud to himself.

A short walk around the station enabled the young American to take in as much of the city as he could on his first exposure. Not a moment passed without some sign of the military or the presence of war. While Edinburgh was by no means a besieged city, it was clearly prepared for what many were convinced was about to come. The Army and Air Force uniforms did not outnumber the Navy uniforms. Brian took great pleasure in the calm, warmth of the people among the beautiful trappings of the city. His enjoyment nearly caused him to miss the train departing to the East out of the city.

The tracks took him away from the hills of Edinburgh along the southern shore of the Firth of Forth. The low overcast and drizzle made the water appear ice cold and foreboding. The gentle rolling hills of the coastal plain were covered by patches of green and brown. Farm country seemed to be the same the world over. The most noticeable differences were the borders of stone instead of hedgerows of England and the straight dirt roads of Kansas, and the fact that, like England to the South, nothing appeared to be straight. Everything curved and twisted.

The rail trip to Drem took thirty minutes. The small station platform could not match the length of the train. Brian stepped down onto the gravel apron of the rail line and walked the short distance to the small station. Only one man serviced the station.

"Can you tell me the way to RAF Drem?"

"Most assuredly I can, young man. You take this lane here up the hill to the third left. You then go to the lane just after the widow Slaughter's house. Follow that carriageway out until you see the aerodrome."

"How far is it?"

"About three miles, I should think."

The thought of a three mile walk on this damp, chilly autumn Saturday did not particularly appeal to Pilot Officer Brian Drummond. "Can I get a taxi?"

The station master laughed at the ridiculousness of the question. "Taxi, my boy? This is a tiny country village. There are no taxis."

Feeling a strange combination of embarrassment and resentment, Brian set out from the station.

"You could wait until one of the RAF lorries comes by to retrieve their supplies," the station master called out to Brian who was ten meters away. Brian ignored the man and resigned himself to the soggy walk.

The countryside would have been more beautiful as he walked if he had not begun to feel uncomfortable. The wetness reached through his uniform about the time he reached the quaint stone house with a carved wooden sign, SLAUGHTER, by the front gate. The chill knifed through his body tempting him to seek shelter and a hot cup of tea with the widow Slaughter. The thought of intruding upon someone else kept him from the relief.

Near the top of the first rise from the Slaughter house, Brian heard an automobile or truck coming up behind him. It was clearly a military supply truck. Maybe he would be able to get a ride? He waved to the driver signaling him to stop which clearly did not make him happy on the slope of the hill.

The driver rolled down the window. "May I help you, sir?"

"Can I get a lift to the aerodrome?"

"Certainly, sir."

Brian hefted his luggage case into the nearly full bed. As he closed the door behind him, Brian immediately felt the warmth of the interior and shuddered with the contrast.

"You are the new American, aren't you, sir?"

"Yes."

"Wasn't anyone at the station to meet you?"

"No."

"The Sergeant Major will not be happy," he said more to himself than to Brian.

Several miles down the narrow country lane Brian could see several large fabricated buildings that had to be maintenance hangars for the aircraft. Just as the distinctive shape of the gate security shack came into view, three fighters accelerating at high power passed low overhead the lorry. Brian instinctively looked across the lorry cab to see the flight of Spitfires turning and climbing back to the East.

"God, I love the sound of those engines," Brian stated proudly.

"The Merlins do make a right pretty tone, don't they?"

"The best."

Armed guards and a barrier protected the entrance to the airfield. A small sign proclaimed the facility, Royal Air Force Drem Aerodrome. The guards checked the cargo bed of the truck and Brian's papers before raising the barrier. The driver stopped in front of the small wooden building identified as, No.609 (West Riding of Yorkshire) Squadron Operations.

The administrative procedures of joining were the same and were becoming routine. The novelty of his citizenship remained the most dominant topic of discussion for new relationships and contacts. Brian knew this check-in process was different. This was his first operational squadron. He expected to remain with this group of men for the duration of the war. His relationships within the unit would affect his performance as Malcolm Bainbridge had told him so many times prior to leaving Kansas.

"We did not expect you until Monday a week, sir," said the clerk. "We would have picked you up at the rail station. The Sergeant Major will be most displeased with us."

"I've heard that. My apologies for causing any disruption."

"It is not your fault, sir." The clerk collected up all the proper papers, sorted through them one last time, then placed them in the appropriate places. "I believe that is all we need. I know the Squadron Leader will want to see you. Unfortunately, he won't return until later today. In the meantime, the Officer's Mess is up the road and to the right, sir."

Pilot Officer Brian Drummond found the Officer's Quarters, an old stone farmhouse with wooden additions for more space, virtually deserted except for the mess stewards. His assigned room had his rank and name already affixed to a board next to the door. The simple room of bare walls, a small window with simple curtains, a modest standing dresser, a typically small bed that he was gradually becoming accustomed to, and a writing desk with two chairs. Not a particularly elegant room like the Savoy Hotel in London, but it was his. The first individual room he had since joining the RAF.

Brian unpacked his case hanging most of his clothes in the standing dresser. The damp articles including those he was wearing were spread out across the furniture to dry. A long hot shower abolished the last remnants of chill from his body. With nothing else to do until supper, Brian decided to take a walking tour of the small airfield facilities. The rain had stopped, replaced with a brisk wind.

"Dinner has passed but would you care for a sandwich and a cup of tea, sir?" asked one of the stewards.

"That would be greatly appreciated."

The task took just a few minutes. "You are the new American assigned to the squadron, if I might ask, sir?"

"Yes, I am."

"Good to have you with us, sir. You'll like the unit. Top rung, if I do say so."

"Thank you."

The roast beef sandwich and hot cup of tea with milk and sugar knocked the edge off his hunger. After conveying his appreciation and trading a few stories with the stewards, Brian took a long walk in the blustery Scottish afternoon. Everything was green, even some of the buildings where the endemic moisture encouraged the moss. He spent the most time along the flight line and in the hangars absorbing details of the Vickers Supermarine Spitfire.

The cockpit instrumentation and controls were similar to the Gladiator. The characteristic circle on top of the control stick, the pilot's called it the spade, dominated the small compartment. The ragged remains of the gun port tape and the heavy gun soot created eight distinctive smears along the top and bottom of the wing. The smell of burnt gun powder among the usual odors of oil and gasoline added to the scene. Brian smiled to himself. He was finally in a business squadron. These guns had been fired in anger. This was a real live working squadron of fighters. He had to resist the temptation to strap on one of the ready machines and go fly. His patience strained under the temptation.

The growing darkness signaled the end of the flying day. One flight, three pilots, the patrol that had taken off as he arrived, and a standby flight were all the attendees at evening meal. Brian was introduced to half of his squadron mates. The vigor, humor and lightheartedness of the pilots impressed the rookie. He felt welcome and part of the team although he could not keep all the names straight, yet.

Most of the pilots departed rather promptly after the meal. Listening to the banter, the Scottish lasses were quite similar to their Southern sisters in being affected by the men who flew. Brian considered going into Edinburgh with the others, but decided to remain behind in anticipation of meeting his commander.

The quiet time sufficed for letter writing to his parents, Malcolm Bainbridge and especially Group Captain John Spencer thanking him for his assistance in being assigned to a Spitfire squadron although he did not have confirmation. He also struggled with the decision to write to Rebecca Seward. In the end, neither Becky nor Anne would get a letter. The distance from his school sweetheart continued to widen as his last waking thoughts were of Anne Booth.

Chapter 14

> One machine can do the work of fifty ordinary men.
> No machine can do the work of one extraordinary man.
> -- Elbert Hubbard

Monday, 23.October.1939
RAF Drem
Drem, Lothian, Scotland

P<small>ILOT</small> Officer Brian Drummond walked the distance from the Officer's Residence to the No.609 Squadron Dispersal building in the early morning, chilly twilight. Breakfast did not appeal to his interest against meeting his new commander and potentially flying his first flight in the Spitfire. Only one person occupied the pilot's gathering place, a moderately attractive, female, Women Auxiliary Air Force, corporal. The short, brown, curly hair contrasted pleasantly with the WAAF uniform.

"Morning, sir. May I help you?"

"I'm Brian Drummond. I'm the new pilot."

"Please to meet you, sir. I am Corporal Jennifer Warren, the squadron operations clark. Nice to have you with us. I imagine you shall want to see Squadron Leader Darling."

"Yes."

"He should be along shortly. Would you care for a cup of tea?"

"That would be nice. Thank you."

Corporal Warren answered Brian's questions as she described how she did her job. The squadron tote board occupied the most time. The white chalk status of each aircraft as well as each pilot filled spaces on the partitioned blackboard. Brian did not see his name on the board, yet. The WAAF corporal displayed a keen wit and robust sense of humor as well as a confident grasp of squadron operations.

A medium height and build man with plain features and well-trimmed, light brown hair walked into the Dispersal building. The uniform of a RAF squadron leader meant he was probably the commander.

"Skipper, here is our new throttle jockey, Pilot Officer Brian Drummond, a week early," announced Corporal Jennifer Warren. "Mister Drummond, this is Squadron Leader Darling."

Darling extended his hand. "Brian, is it?"

"Yes, sir."

"Welcome to our little band of merry men," said Squadron Leader Horatio Darling. The rugged features of the smaller man added to his presence.

"Are you ready to fly?"

"Yes, sir," Brian answered with undeniable enthusiasm. "I've been waiting a long time for this."

"Have you now? Your record indicates you flew Gladiators in the Training Command."

"Yes, sir."

"You will find the Spit has significantly more performance. Only two things you need to watch out for. One, the undercarriage is rather narrow. Take care with your landings and especially with any crosswind. Second, the Merlin is a normally aspirated, float-type, carburetor engine that will cough if you go light on the 'g.' Just keep the wing loaded up. Questions?"

"No, sir."

"Well then, lad, I should say we ought to give it a go. So, snatch your kit and let us go cut some sky."

The lack of any reference to his nationality did not escape Brian's notice. He appreciated the relief from the incessant questions and the implicit acceptance of his membership in the fighter squadron. Brian just wanted to be a co-equal part of the unit and enjoy the flying.

As they stepped out of Dispersal, Brian took a quick look to the sky. The solid, nearly uniform overcast on the fall day meant the cloud layer was not particularly thick. They would have to do a penetration departure and return. A light breeze blew from the west out of the mountains and into the North Sea. Not a great day to fly but not a bad day either.

Walking toward the flight line Darling said, "As you know, we usually fly in flights of three. This will be a training flight, Mister Drummond, so only the two of us will take this one. If your records and reputation are correct, this will be your one and only training flight. We will be fully armed. You shall have the lead. If Gerry decides to join the party, I shall have the lead. You will need to fly a good wing position. Keep your eyes out every instant of the time. Your eyes are your survival. Clear?"

"Yes, sir."

"If my sources are correct, you went by the callsign, 'BAD,' in the Training Command."

"Yes, sir."

"Why was that, might I ask?"

"My initials are, BAD."

"Clever. The squadron callsign is, 'Sorbo.' My callsign is, 'Spike,' and I do not have a clue how I inherited it."

"Got it, sir."

"You will have the 'Freddie' bird. She is spanking new, just delivered from the Woolston factory a few days ago. I will be in 'Ack.' One more thing, I hope they told you not use, sir, or any other words of authority in the air."

A nod of his head seemed to be the appropriate response.

The tail of each aircraft had the large white letters, PR-F, the code letters for No.609 Squadron, PR, painted just aft of the cockpit on one side of the red, white, blue and gold roundel. The unique aircraft code letter occupied the other side of the roundel, in Brian's case the letter, F. The squadron leader's aircraft would have the letter, A.

Brian found his aircraft. The flat green and brown camouflage colors were fresh and new. The manufacturer's number, MN4011, was painted in small black letters just forward of the tail.

"What maneuvers do you want?" asked Brian.

"You just do what I tell you to. I want to see how you think. Any questions?"

"No, sir."

"Here we are. This is your machine. Gordon," Darling called out.

A short stocky man in fitter's coveralls and corporal stripes on his arms left the aircraft. As he approached the two officers, he rendered a snappy, palm-out salute, which was returned by both pilots.

Darling turned to Brian. "This is your crew chief, Leading Aircraftman Bernard Gordon. Bernie, this is your new pilot, Pilot Officer Drummond."

"Pleased to meet you, sir," the leading aircraftman said.

"It's great to see you, Leading Aircraftman Gordon," said Brian with uncontained excitement. He knew he was getting closer to his dream.

"If you will permit me, sir, I'll introduce your crew. Jordan," Gordon called out. A thin, average height pimply young man came around the nose of the aircraft. "This is Aircraftman Jordan Toldson, our rigger. The bloke under the wing," a rather plain, rough looking, small man stopped his work with the guns, "is Aircraftman Colin Jenkins, our armorer. Best crew in the squadron, I'd say."

"It's great to meet you all," Brian said shaking hands with each of his crew.

Bernie Gordon waved his hand around as if he was introducing a distinguished, guest of honor. "If you will keep just one thing in mind, Mister Drummond, she's my machine. I would be honored to have you fly her for us, but I must ask you . . . do not bust her," added the crew chief.

"Make sure you listen to the chief," said the squadron leader over his shoulder as he continued on to his aircraft.

"I'll do my best," Brian said to his crew with a broad smile. "I've been waiting months for this flight. I'll take care of your machine."

Brian zipped up his leather jacket, donned his seat pack parachute and stepped up on to the left wing. Brian settled into the small cockpit. Leading Aircraftman Gordon assisted Brian with the harness straps. The leather helmet with integral earphones slipped over his head like a glove. The connections for his headset and oxygen mask clicked into place. Leather gloves completed the cockpit attire.

The cockpit looked different this time, more real and alive. The dark gray of the panel with the black round dials of the instruments with their white array of numbers and pointing needle arms provided the means of communication with the machine. Like a mute, the feelings of the machine had to come through the panel of dials. Brian's partnership with the Spitfire demanded his attention to the principal method of communication as well as the delightful, but meaningful hums and throaty tones of the big Merlin engine in front of him.

Preparations for engine start and flight by both men began with calm precision. The trim wheels and flight controls were run through their complete travel to ensure there was no blockage or binding. Brian checked everything twice to make certain he missed nothing. Leading Aircraftman Gordon closed the access hatch, locked it and ran his hand over the fuselage skin to confirm the edges flush with the fuselage.

Brian sucked in a deep breath, checked both sides of the propeller arc and shouted, "Clear."

"Clear," Gordon answered to confirm the propeller was free of obstructions and Brian was allowed to start the engine.

Brian pushed the primer button five full times to squirt fuel into the intake manifold and screwed the primer down to avoid inadvertent injection. Ignition switch, ON. Throttle lever opened about a ¼ inch. With one last quick check, Brian depressed the Type L.4 Coffman starter button and heard the cartridge fire. The enormous, two bladed, fixed pitch, wooden propeller turned through three blade passes before the engine fired. The cloud of black smoke and the biting odor of partially burnt gasoline belched from both exhaust banks as the big engine rocked the aircraft. All twelve cylinders were soon firing smoothly.

Brian switched on the T.R.9D radio ensuring the transmit & receive switch was in the RECEIVE position. The attitude gyro was uncaged and erected, and the compass aligned before he looked down the line of Spitfires to give his leader a thumbs up that he was ready.

With a return sign, Pilot Officer Brian Drummond crossed his arms in front of him motioning with his thumbs out for Leading Aircraftman Gordon to remove the wheel chocks. The engine responded to the nudge of the throttle moving the aircraft gradually forward.

Once clear of the parked aircraft, Brian pushed the right rudder pedal in and advanced the throttle slightly to turn right. Darling was already ahead of him snaking his aircraft to look past the broad nose to ensure nothing was in front of him. Brian followed suit.

The propeller wash made the cool, moist air more chilly. The fresh feel to it enticed Brian to suck in another deep breath and feel the cold in his chest. At the downwind end of the airfield, two fitters in their winter overalls waited for the two Spitfires. Darling lined up first and began his pre-takeoff checks. Brian turned to position his aircraft beside the left wing of his leader.

Adding power to run the engine at a fast idle, Brian checked the coolant temperature. 80° . . . well below the 130° limit . . . plenty of room for the engine checks. He waited his turn.

The sound of Darling's engine at full power filled his covered ears. As he throttled back, the ground crewmen left his tail to tend to Brian. The canopy closed. The tail moved as the two men leaned over the leading edge of the horizontal tail to hold it down for the high power engine check.

Engine temperatures and pressures within limits. Brian applied full brake pressure, pulled the control stick full back into his lap and pushed the throttle forward up to the emergency gate, actually just a strand of frangible wire across the slot, adding left rudder to compensate for the strong torque of the powerful engine. Roaring to full power, the engine rose to 2800 rpm with the boost gauge indicating +3 psi. Oil pressure OK. Just right, Brian told himself as he retarded the throttle to 2000 rpm, reached for the magneto switch, moved it from BOTH to L, rpm drop, 50 rpm, back to BOTH, and then to R resulting in a drop of 40 rpm. Pull throttle back to hold 1200 rpm. All gauges normal. Throttle back to idle. Brian watched the two men move away from the two fighters, looked over to Darling and gave him a thumbs up. Darling tapped the edge of his open canopy to remind Brian to open his canopy for takeoff just in case he had to get out quickly. In turn, the squadron leader motioned to the new pilot that he had the lead and go ahead with takeoff.

"Drem Tower, Sorbo Blue, two for takeoff," Brian called on the radio.

"Sorbo Blue, winds two eight zero at seven, you are cleared for takeoff."

The head nod from Darling acknowledged the call. Brian released the brakes and slowly pushed the throttle forward. The aircraft accelerated gradually in the soft grass. Picking up some speed, the rudder became more effective, and Brian added more power. As the tail rose off the ground, he adjusted the controls to hold the tail level and the nose straight. The sprightly Spitfire bumped and shuddered as the wheels passed over the slightly uneven ground. Soon, the Merlin was at full power. The main wheels became light and springy as the Spitfire lifted into the air accelerating smartly past 100 mph. Brian switched hands on the control spade to move the undercarriage lever to the UP position and pumped the handle to raise the landing gear. Despite the warnings from other Spitfire pilots, Brian banged his knuckles hard on the nearby frame. The clunk of the uplocks established the landing gear was fully retracted. Brian closed and locked his canopy. The full speed range was now available.

The Spitfire continued to accelerate as Brian banked right, turning to the North. The slight motion of Darling's Spitfire off his right wing attracted his attention and gave the only sign he needed that everything was OK. Quickly scanning his instruments, the message from the airplane was the same. Everything was right with the world and the broad grin on Brian's face complemented the feelings. This was better than he could have ever imagined.

"Rooker Control, Sorbo Blue, with you passing Angels Two," Darling called out on the radio.

"Roger, Sorbo Blue, understand training flight."

"Affirmative."

"Roger. No activity. You are clear in the Northern Area. Call upon return."

"Understood."

The two Spitfires entered the cloud layer at 2500 feet. Brian concentrated on his instruments scanning from the artificial horizon indicator to the compass, altimeter and climb indicators to keep the fighter climbing with the wings level. He knew the other aircraft would be close by, but he would not try to see if Darling was there. The clouds began to thin just before Brian's aircraft broke out into the bright blue morning sky. The cloud deck below him stretched as far as he could see in every direction like an enormous white blanket covering the Earth. To the West, the mountains of Scotland jutted up through the blanket.

Brian looked out his right wing to see Darling's PR-A bird glued in perfect position. A quick thumb's up confirmed his status. He returned the signal.

"I am going to back off, 'BAD.' Exercise the machine. Get the feel of her, and then we will play some games."

Brian raised his right thumb to indicate he understood the directions.

He immediately lifted the nose about ten degrees and rolled his Spitfire away from Darling who was already a good distance back. Several more rolls followed. Only one caused the engine to miss a few beats as he allowed the aircraft to get a little light cutting off the fuel momentarily to the engine. Brian kept up a continuous series of aerobatic maneuvers, each one giving him a feel for the forces on the controls, the acceleration and deceleration associated with each maneuver, and power required.

The aircraft talked to him in the common language of tones, hisses, groans, creeks and whines. Brian knew he needed to learn the subtleties of his aircraft's language in order to know where her limits were. The control forces were quite light over a wide speed range that meant the aircraft could be handled easily in a chase. The tracking capability was smooth, uniform and predictable. This was definitely a hunting machine, Brian told himself.

Looking around the sky, there were no other aircraft in sight. Suspicion brought the young fighter pilot to a razor sharp edge. Brian knew what was coming. He searched the sky above him and toward the Sun. The brief glint of Darling's canopy told him what he needed to know. The squadron leader was positioning for a mock attack from out of the Sun.

Brian pushed the throttle full forward. The engine groaned and responded. The boost gauge needle sprang to +3. Brian pulled back on the stick and eased the airspeed back to best climb speed. Straining for altitude, Brian kept looking for Darling.

Catching the distinctive shape of the Spitfire wing, the attack was on, although not yet announced. Brian turned into the Sun climbing directly into his diving leader.

"Tat-tat-tat-tat-tat," came the pretend machine gun fire over the radio.

The blur of the diving Spitfire passed to his left.

Brian rolled sharply left, banging his head on the close canopy, and pulled the nose of his Spitfire hard. He tightened his stomach muscles, grunting hard to fight against the looming blackout of high g's as blood was forced from his head. The aircraft shook and shuddered near maneuvering stall but responded as commanded. Brian craned his neck to regain sight of his pretend adversary. Darling was climbing and turning to return to the attack. Brian had the advantage this time. He maneuvered to maintain the advantage as they closed the distance.

Soon, the two Spitfires were entwined in an aerial ballet familiar to all fighter pilots. The predecessors in the Great War coined the term, dogfight, since it resembled the natural occurrence. The speeds of modern fights made the dimensions of the dogfight greater but the characteristics remained the same.

Several more engagements from different starting positions were performed. Brian held his own against the more experienced fighter pilot losing slightly more than he won and drawing to a standoff on most of the encounters.

"How is your petrol?" asked Darling after eighty minutes of mock combat.

"About a third remaining."

"I have the same. I have seen what I need. Let's head for home. I have the lead."

The cloud deck was still solid except for the high mountains of Central Scotland that were now below them. Darling headed to the east wanting to get over the North Sea before letting down through the clouds. Brian maneuvered to take a position just behind and off Darling's right wing. His attention narrowed to his leader, matching each movement. He promptly began searching for and defining his formation position cues using the features of the lead aircraft as well as his canopy.

After the appropriate radio calls, the two Spitfires descended. Brian moved a little closer to the lead aircraft as they entered the clouds. The rushing movement of clouds brought the usual distraction and twinges of disorientation. Darling flew an excellent lead keeping his aircraft steady trying to minimize the changes, to make Brian's task less difficult. With a couple of openings flashing past them, they broke out under the cloud deck. They were indeed over the North Sea. Land was not in sight. Darling turned back to the west. Within five minutes, the jagged rocky coast of Scotland appeared out of the haze, causing Darling to turn to the south. In another fifteen minutes, the coastline fell away into the Firth of Forth estuary.

"Drem Tower, Sorbo Blue, two fighters north for landing."

"Sorbo Blue, Drem Tower, no traffic, wind two eight five at eight, cleared to land."

Brian changed hands on the stick placing his right hand on the undercarriage lever. He waited until he saw the landing gear on Darling's airplane begin to move, then he moved the handle to lower his own wheels. The roar of the air passing over and around the extending main landing gear as well as the jostling of the aircraft confirmed his undercarriage extension. Brian hit his knuckles one more time. He was certain his knuckles had to be bleeding. He quickly moved the controls to adjust for the jostling of aircraft until the gear clunked into the locked, down position. Brian checked that the undercarriage downlock pins extended from the upper wing surface as positive mechanical confirmation his wheels were down and locked in place. Likewise, he used his leader to cue him when to lower the landing flaps. Brian worked hard to maintain a perfect position on Darling's right wing. The wind was straight on the nose, no crosswind. Brian's aircraft touched down first since his position was slightly lower than Darling. As they taxied back to the line, Brian spotted Leading Aircraftman Gordon. He followed Darling behind the line of Spitfires, and then turned into the space with Gordon in the middle. Moving forward until his wing tips lined up with the other aircraft, Brian watched Gordon cross his arms signaling him to stop.

Checking the gauges quickly to ensure everything was as it should be, Brian closed the throttle and switched the ignition OFF. The large wooden blades stopped quickly. Brian pulled the canopy open to feel the cool, fresh air. As he unstrapped, Gordon opened the access door.

"How did she run, sir?"

"Like the precision machine she is, chief."

"Good to hear it. Any snags?"

"None."

"Equally as good, then. Are you and the skipper going back up?" The implicit element of the question portrayed what was probably the most common scenario where a new pilot did not quite get it right and needed to go back up for more practice and instruction.

"I don't think, so. We'll need to ask the squadron leader."

"If he didn't say it, you probably nailed it, I'd say."

Brian waited by the tail of his airplane. Darling took some time to talk to his crew chief. Maybe he had some problems with his aircraft.

"It would appear, Mister Drummond, you are up to your reputation. Excellent flight."

"Thank you, sir."

"Going back up, skipper?" asked Leading Aircraftman Gordon.

"No. We are done."

"Super," Gordon said with a broad smile probably reflecting his own pride in his new pilot. Each crew chief wanted his pilot to be the best, to have the most kills with the least damage. Gordon undoubtedly felt he finally had a good candidate. "The machine will be ready when you are, Mister Drummond."

"Thank you, Bernie," Brian said over his shoulder as Squadron Leader Darling began walking back to the Dispersal hut. He also took another long look at the sleek, smooth lines of his aircraft. The thought brought a smile to his face. He liked it.

Squadron Leader Darling's brisk pace in the early morning light made

Brian run to catch up. "Your performance with the Spit was quite admirable for the first go, Mister Drummond. You are to be commended," Darling pronounced.

"Thank you, sir."

"We shall take a more serious step tomorrow. Do you have any questions?"

"What should I prepare for?"

"Make sure you know your Pilot's Notes, backwards and forwards as well as inside out. You might want to talk to some of the other lads about the machine and their techniques for flying it. You should also fill in some of the details of our operating procedures especially the commands from Sector Control and our methods internal to the flight. We will take up two sections tomorrow."

"No problem. What about gunnery?"

"Your training record indicates exceptional skill in aerial gunnery. While the Spitfire is a substantially greater fighter than the Gladiator, the principles are the same. As I am certain you are aware, we have a rather ominous storm in front of us. We will do some gunnery, but we will not do enough. We must preserve our ammunition for Germans. We must rely on your skills."

Brian considered the words and the image they created. The thought of going into combat against a determined adversary flying Willie Messerschmitt's vaunted Bf109 without having practiced with his own guns did not give him any comfort.

"Will I fly your wing?" asked Brian.

"No, thanks for asking. As I said earlier, you will fly number three in Green Section, left wing, with Flight Lieutenant 'Jackstay' Beamish. Any other questions?"

"No, sir. I suppose not."

"Then, I shall see you in the Mess."

Thursday, 26.October.1939
RAF Drem
Drem, Lothian, Scotland

"SCRAMBLE the squadron," came the command from Squadron Leader Darling at mid-morning.

All the pilots grabbed their flying kit: leather helmet with earphones, oxygen mask, goggles, gloves, Mae West floatation vest, and ran toward the line of Spitfires waiting for them. The ground crews were either at or within yards of their aircraft and ahead of the pilots.

In less than a minute, the air filled with the gray smoke from 12 Merlin engines as the fighters of No.609 Squadron came to life. Darling's aircraft was the first to move. Brian looked for the proper opening to follow his section leader, Flight Lieutenant 'Jackstay' Beamish, in the PR-D, 'Don,' aircraft. He completed strapping himself into the aircraft and closing up the cockpit as he moved toward the takeoff point at the downwind end of the airfield.

In another minute, the powerful melody of engines quickened 'BAD's heart rate. He waited for 'Jackstay' to move before he throttled up the Merlin. The three aircraft of Green Section bumped and bounced across the takeoff area and were soon in the air. The landing gear came up immediately. 'Jackstay' turned the section to follow 'Spike's Blue Section climbing to the east into the North Sea. The section passed through a relatively thin overcast layer. Once above the clouds with bright blue sky drawing them up, Brian charged his eight Browning 0.303 machine guns in preparation for combat. Quickly, he checked all his instruments and switches to ensure everything was ready.

"Sorbo, this is Rooker. Climb to angels one five, vector zero seven seven. Negative contact," the sector controller radioed for the flight.

"Roger, Rooker," 'Spike' answered as he adjusted his heading to 077° magnetic.

'BAD' pulled the throttle back to show +1 psi of boost, then adjusted his power to stay in position on 'Jackstay's left wing. Red and Yellow Sections joined up with section 'V' formations in a right echelon. 'BAD' concentrated on flying his position with his right wingtip moving gracefully less than ten feet from his leader's left wing. He was virtually in the middle of the most awesome array of high performance aircraft he had ever seen. The temptation to become a spectator was difficult to resist, but this was serious business. 'BAD' did not have the familiarity he needed to be comfortable with his machine, yet. Although he did not have the time to know all the subtleties of the aircraft, but like the other pilots, he relied upon his recognition of the sounds of the Merlin engine as well as the air rushing passed the canopy to establish the health of the aircraft without looking at his instruments. Even the commands from 'Spike' to change formations, or from the sector controller to redirect the squadron did not break 'BAD's attention to the proximity of the other aircraft around him. He had no way to know where they were, or actually what altitude they were at any point in time.

'Spike' took the squadron through a full series of practice attacks with various formations and tactics. Practice was orderly and purposeful. With about half their fuel load used, 'Spike' headed the squadron back toward the southwest. The clouds below filled the sky like a sea of lumpy vanilla ice cream. To the west, peaks of the Scottish Highlands were quite far off. As they descended toward the cloud deck, 'Spike' commanded the squadron into section V's in a line astern. As they entered the clouds, 'BAD' moved closer to 'Jackstay' to ensure he did not lose sight of his leader in the low visibility of the clouds. The cloud deck seemed to be several thousand feet thick that did not take long to penetrate. As they broke out below the clouds, 'BAD' backed away from 'Jackstay' to orient himself. They were over the North Sea with no land in sight. Soon, they changed heading to 280° that brought the coastline of Scotland into view within several minutes.

'BAD' watched 'Jackstay' with the focused concentration he had been taught over the last few months. As they approached, 'BAD' knew 'Spike' had lined them up for a straight in landing at Drem. He prepared himself for the landing checks. 'Jackstay's undercarriage began to move out of the wing pockets. 'BAD' quickly changed hands on the stick to lower his wheels. 'BAD' waited for 'Jackstay' to lower landing flaps before he lowered his. A quick scan of the cockpit confirmed everything was ready for landing which occurred uneventfully.

"How was your first squadron sortie, Mister Drummond?" asked Leading Aircraftman Gordon as Brian pulled the canopy back, opened the access hatch and unstrapped.

The impressions, excitement and energy instantly refilled his thoughts. "It was glorious. Hard work but absolutely glorious."

"Good to know you enjoyed yourself, sir."

"Thanks."

"I will have her checked and refueled," Gordon paused to see the wing gunport patches still in place. "No need to rearm. I'll have her ready in a flash."

"Thank you, Bernie. No snags with the bird."

"Good. I will do my best to keep it that way."

Pilot Officer Brian 'BAD' Drummond joined several other pilots whose names he could not yet remember for the walk back to Dispersal. Squadron Leader 'Spike' Darling debriefed the flight which had gone quite well although, as Darling said, there was always room to improve. With his feet firmly on the ground and without another aircraft floating close by, Brian listened to Darling's words and renewed his concern for the vulnerability of close formation flight. The words of Malcolm Bainbridge returned to him with great clarity. The aerial combat lessons he learned from his tutor, mentor and friend were almost entirely contrary to what he was now flying. The proximity of other aircraft meant his attention had to be focused on his leader almost continuously until the dive for an attack. How was he to watch for enemy fighters or even spot lumbering bombers when he was so close to his leader? The questions were pushed to the back of his consciousness. This was the Fighter Command of the Royal Air Force. They were flying the best fighters in the world. The tactics had to match the organization and the machines. Brian convinced himself he was simply not experienced enough to understand the significance and subtlety of the tactics. He knew in time he would understand. He simply had to be patient.

The squadron intelligence officer, Flying Officer James Royster, joined the group in Dispersal. He debriefed each pilot individually. The mission was uneventful and primarily a training event. Royster spent the time with Brian explaining how the debriefing process worked and tried to impress Brian with the need to retain an accurate image of a combat flight with as much detail as possible. He was the last pilot interviewed by Flying Officer Royster, a non-pilot, three-year professional RAF intelligence officer. Most of the other pilots had already retired to the Mess.

As Brian entered the Officer's Mess, a humble little, rough-hewn, clapboard shack adjacent to their quarters, no one seemed to recognize him, at first. Then, the familiar face, although he could not remember his name, of the right wing of his section lit up and came forward. "You probably don't remember among all the other names . . . Steve 'Mongo' Strickland," he said extending his right hand. Strickland's brown wavy hair with bright green eyes and chiseled facial features made him appear to be quite the rouge. He had a strong confident manner and ebullient personality.

"Sure."

"Good piece of flying for a newbie."

"Thanks."

Strickland raised a finger, and then grabbed a pint of beer thrusting it into Brian's hand. "You need one of these to prime the pump," he said. "Where did you take your training?"

"OTU Seven."

"Oh really. I thought they only had a few Spitfires."

"That's correct. I flew Gladiators."

"Well, now, you must know someone."

The reality of his connections to Group Captain Spencer did not add to Brian's sense of fairness. He was immensely grateful for the influence that brought his dream to fruition, but the recognition that he probably could not have done it on his own detracted from his gratitude. "Why does everyone say that?" he spoke more strongly than he wanted.

"No need to be indignant, mate. You should be thankful you do have a benefactor."

Brian felt a twinge of regret with a smear of embarrassment. "I'm sorry, Steve. It's just that everyone thinks I couldn't have done this on my own, and occasionally it does get to me." Brian thought for the first time, this is just a waypoint on my journey . . . just leave me alone.

"Don't get your knickers in a twist, my young Yank. It is most certainly an understood commonality."

"Sure," Brian responded with a more subdued tone.

Several of the other pilots began to talk about their frustration with the lack of action. The only combat seemed to be with the Royal Navy or Coastal Command, the coastline patrol squadrons. The words were strong. Fighter pilots wanted action; they needed action to keep their skills honed.

"If I may have your attention, gentlemen," 'Spike' Darling said authoritatively. "This is the first occasion we've had with the whole squadron present to welcome our newest pilot. I might add for the consumption of our distinguished clan, young Mister Brian Drummond, callsign 'BAD,' comes to us from the colonies across the Atlantic. According to RAF records, Brian is the first American volunteer to join our band of merry men in our present row with the heathen Hun. So, let's welcome Pilot Officer Brian Drummond properly."

At that instant, everyone tossed their beer at him drenching his uniform. Brian was staggered not knowing what to do, or how to react.

"You are properly christened, Brian. It is a squadron tradition much like a bottle of champagne across the bow to launch a ship."

Brian jerked a tablecloth off a table without dropping any of the items on the floor, and wiped his face and uniform. The hoots and hollers from the other officers conveyed their acceptance of his move. The bond between them grew a little more that evening.

The fighter pilot's frustration was the dichotomy of the war. They had been prepared to fight; they wanted to test and demonstrate their skills; and yet, an aerial engagement with the combat-tested *Luftwaffe* was daunting at best. For the first time, Brian heard the term, Phony War. The mood among the RAF reflected their frustrations. Everyone told them there was a war on with Germany, but No.609 Squadron, as with most of Fighter Command, had not seen results from their few engagements so far. The realities of the Phony War made all of them feel that the Prime Minister would someday announce to the nation: all was forgiven; Germany apologizes for raping Poland; the borders had returned to pre-September positions; and, the war was over. There were a few like Squadron Leader Horatio 'Spike' Darling and Group Captain John

Spencer, who felt the looming certainty of a cataclysmic confrontation with Germany, most probably in the skies of Great Britain. They may be in the Phony War now, but it was about to turn all too real, some said. The spring seemed to be the popular consensus among the inevitability believers.

"Word has it, Mister Drummond," Darling spoke directly to him as they all had another pint of beer after supper, "your instructor in America flew with the Royal Flying Corps in the Great War."

"Yes, sir. Four Three Squadron."

"Ah, yes, the Fighting Cocks," Darling said smiling with reminiscence. "As I recall my history, those blokes had a ruddy good row with the bloody Red Baron."

"Malcolm told me his stories about Baron von Richthofen." Brian noticed immediately the attention the Red Baron attracted even now long after his death in the skies over France and more notably among a group of young RAF fighter pilots. He could only suppose, as Malcolm Bainbridge had taught him, aerial combat skills had to be admired and respected, regardless of the practitioner. Captain Manfred Baron von Richthofen, after all, commanded the respect due a legend with 93 aerial victories.

"What did the bloody Red Baron have to say?" asked 'Mongo' Strickland with a rather sarcastic, back-handed curiosity.

"Ah, Mongo, the bastard's dead. Fell to the guns of a 1½ Strutter from what I hear," quipped Flying Officer George 'Angle' Ashcroft.

"A bit off the mark, 'Angle,'" 'Spike' Darling said with authority. "You got the constructor right, but that's it. *Rittmeister* Manfred *Freiherr* von Richthofen was shot down by Captain Roy Brown of Two Zero Nine Squadron flying a Sopwith F.1 Camel. But, just so you do not get too carried away, there is a reason why the *Luftwaffe*'s Fighter Wing One is designated, *Jagdgeschwader Freiherr* von Richthofen. The Red Baron is to this day the highest scoring ace in any air force including ours. Ninety-three victories, I do believe."

Absolute silence swallowed their thoughts. Brian did not speak. In the end, it was 'Angle' Ashcroft who returned to the discussion first. "Yeah, well, but the soddin' bastard is still dead."

"So, what is the bloody Red Baron's secret to aerial success?" 'Mongo' restated his question looking first to Darling, then Pilot Officer Brian 'BAD' Drummond.

"According to Malcolm, Richthofen believed there were only two traits that make a good fighter pilot, aggressiveness and good marksmanship, in essence, a good hunter."

"Is that so?" said Ashcroft.

"I should say that fairly well summarizes it," Darling responded then turned to Brian. "You said your instructor's name was, Malcolm. English?"

"American, sir. His name is Malcolm Bainbridge, and he was an American volunteer with Four Three Squadron."

"A bit like you, Mister Drummond, but a good English name nonetheless."

"Yes, sir."

"At any rate, we can all learn some good lessons from the Red Baron. We must all be students of our craft, our profession. Remember, he who fails to learn from the past, is doomed to repeat it. Now, let's call it a night, we have alert duty first off tomorrow."

—

Saturday, 28.October.1939
Fighter Command
Bentley Priory
Stanmore, Middlesex, England

GROUP Captain John Spencer did not take any pleasure in calling the late evening meeting of the Fighter Command Staff. Compensation came with the unusual enthusiasm of the intelligence section. He knew something important happened. Anticipation banished the dark of the night. As he waited for the other members of the staff to gather, there was still a resentment that the Germans had turned out the magnificent lights of London. The vantage point of the old monastery overlooking the city always attracted the attention of visitors and residents alike.

"Well, what do you think the shadow masters have for us, John?" asked Air Commodore Herbert Maple, DFC, the Deputy Chief of Operations.

"I suspect they may have found something in that Heinkel One One One shot down earlier today in Scotland. I heard the aircraft was in pretty good shape considering the forced landing."

"What could it be? A bomber is just a bomber."

"We shall find out soon enough, I suppose."

The other officers gathered quickly followed promptly by Air Chief Marshall Dowding. "Good evening, gentlemen," he said in his usual stern, subdued voice.

"Good evening, sir," the staff officers answered in unison.

"What have you got?" Dowding asked directly to his Air Intelligence Officer.

Air Commodore James Hogan, DSO, Chief of Air Intelligence stood. "As you know, sir, we had a Heinkel One One One shot down in Lammermuir

Hills in Scotland. The aircraft is in unusually good shape given combat and forced landing. What is unique about this machine is the electronic equipment fitted in this aircraft. We managed to prevent the surviving crew from destroying it, which they were quite clearly eager to accomplish.

"First assessment of this equipment along with other source data indicates this may be what the *Luftwaffe* documents refer to as, *Knickebein*, or bent knee. We believe this is a night or foul weather electronic bombing aid. The principle, as we surmise, enables a bomber fitted with this kit to bomb a target without visual contact."

"Damn Krauts. What are the bastards going to think of next?" came muffled comment from one of the other officers.

"Quite," continued Hogan. "The equipment is serviceable as far as we can tell. We will extract the kit, install it in a Hudson, and do our best to establish the modes of operation and hopefully what countermeasures we might be able to develop."

Dowding raised his hand, and then looked to Air Commodore Maple. "Herb, what is our status on the night fighters?"

"The Fighter Intercept Unit is working hard to build a night engagement capability, but they have had little success."

"We must obtain a night intercept capability, and of course, we need some form of countermeasure against this *Knickebein* equipment. John, I'd like you to keep a close eye on the evaluation process to feed the data to the Fighter Intercept Unit to aid in their night work."

"Yes, sir," answered Group Captain Spencer.

"Do you have an idea how this *Knickebein* kit works?" asked Air Commodore Maple.

"Actually, not entirely. We believe it operates with two narrow, focused radio beams aimed to intersect over the bomb aiming point for the desired target. The bomber then flies down one beam until it intersects the other beam. When the signal conditions meet the prescribed parameters, the bombardier gets a release light."

"How long will it take to confirm the mode of operation?" asked Dowding.

"A few days to a few weeks, sir. We don't have enough information about the radio beams. We need the 'Y' Section lads to get up and look for the radio beams to give us the details. There may be several subtleties to this equipment. We'd prefer to make sure we know as much as possible before we draw any conclusions."

"Understood, but the thought of night precision bombing over Britain without a countermeasure or a night intercept capability will not go down well with the War Cabinet. We must find a stopper, or we shall be in for a rough ride. Have you interrogated the crew?"

"We have had a chat with them. Unfortunately, the bombardier was killed in the engagement. He may have been the only knowledgeable member of the crew."

"Most unfortunate."

"Yes, sir."

"Anything else?"

"Yes, sir. We have strong indications from various sources the Germans have flown a turbine-powered aeroplane. The aircraft is designated Heinkel One Seven Eight. First flight was 25th of August, this year. There is also evidence the Messerschmitt works are busy on their own turbine-powered aircraft."

Silence filled the room as everyone digested the startling information. John Spencer wondered about the source. It had to be of the most secret level. Did they have an agent inside the German aviation industry, or maybe the *Luftwaffe*? He also knew the implications of the information. Frank Whittle's jet engine research at Brownsover Hall was at least a year behind the Germans. All the calculations pointed toward jet engines pushing fighters to much higher speeds that to any fighter pilot meant substantial advantage. John Spencer wanted to know so much more.

"That is not particularly good news, now is it?" said Dowding.

The group continued to be absorbed with their own thoughts. To John Spencer, the image of facing aerial combat with an adversary having a 50 or 100 mile an hour speed advantage brought a shudder and a twinge of dejection. The feelings passed, replaced by a stiffened resolve to overcome the disadvantage.

Dowding looked to Spencer. "Where is Whittle?"

"About a year behind from the last time I reviewed the program."

"You might want to take another look and see what we can do to help things along. The prospect of facing the *Luftwaffe* with a turbine fighter does not leave one with a comfortable feeling." The pause kept the silence. "You think Whittle might fly next year?"

"At the earliest, sir," responded Group Captain Spencer.

"Who has access to this information?" asked Dowding.

"This is MOST SECRET - SPEED information. Only key personnel within SIS, the Air Ministry and Fighter Command staff have access to these data."

"John, after you visit with Frank Whittle, please establish a close hold project with the Fighter Intercept Unit to work on tactics against higher speed fighters. They must understand and appreciate the sensitivity of this information and must restrict involvement."

"Yes, sir."

"Any other interesting items, James?" asked Dowding.

"No, sir."

"Then, I would suggest we have our work cut out for us. Let us get to it in earnest. Good evening, gentlemen."

Chapter 15

> Fate chooses our relatives,
> we choose our friends.
> -- Jacques Delille

Saturday, 4.November.1939
No.10 Downing Street
Whitehall, London, England

"THANK you for coming, Winston," the Prime Minister said, extending his hand to the First Lord. The door closed leaving the two veteran politicians in the prime minister's office. "How is our situation at sea?" Chamberlain motioned toward the well-cushioned chairs.

"Those damnable U-boats are my greatest worry. Their successes continue to mount while ours are few and far between. The fast convoy system is helping but that will not be sufficient."

"Why?"

"Two reasons. One, there are not enough merchants that can maintain speeds in excess of twelve to fifteen knots. We will have to figure out a way to protect slower ships in the six to eight knot range. We must have greater lift capacity. And second, our offensive tools to hunt the submarines are woefully inadequate. They have the advantage at the moment, I'm afraid."

"Prognosis?"

"Worse before it gets better . . . much worse, I fear."

"What is the latest on the *Graf Spee*?"

"She continues to threaten our sea lanes. She sank several ships in the Western Indian Ocean, but I suspect she did not find sufficient hunting. We believe she has returned to the South Atlantic."

"But, the Navy has not found her?"

"No. We can only locate her victims. We have deployed four battle groups for the hunt. We will find her and sink her. But, after how much damage?"

Chamberlain nodded his head. His thoughts were somewhere else. He knew his questions would only have guesses for answers. He looked back into the focused eyes of his political alter-ego.

"Actually, Winston, I did not ask you here for an Admiralty briefing. You may already be aware . . . I was sadly informed that Sir Hugh passed away this morning."

"No I was not, but he has been ill for several months now."

"Indeed. I need your counsel on his successor as chief of MI6.

Menzies has performed admirably during Sir Hugh's incapacitation. Should he succeed Sir Hugh as 'C?'"

For Winston, the choice was nearly perfect. He could not think of anyone else inside the Secret Intelligence Service, still popularly known as MI6 or the Foreign Intelligence Branch. Colonel Menzies was a long term friend and supporter, but more importantly, an accomplished and respected, professional, intelligence officer, seconded from the Army.

"An excellent choice, Neville."

"Very well. He appears to be the popular and professional choice. He has indicated his willingness to assume the duties. I will notify him immediately and the Cabinet at this evening's meeting." Chamberlain paused without altering his concentration. "Should we promote him to brigadier?"

"He would deserve a promotion from colonel. I would recommend we let the dust settle a little and the appropriate leaders gain more direct experience with him before you entertain a promotion to the general officer ranks."

"Sound advice." Chamberlain took a deep breath. "It is a tragedy we cannot publically recognize the extraordinary contributions of Sir Hugh to the security of the empire."

"Surely, but they know. We know."

"The shadows of darkness."

"We must have it that way," Winston added the obvious.

Again, Chamberlain nodded his head in agreement. The two men concluded their meeting. Winston left No.10 Downing Street with a smile on his face. The smile remained for the short walk back to the Admiralty.

―

Thursday, 7.December.1939
RAF Drem
Drem, Lothian, Scotland

THE warmth of the fire in the wood stove melted the cold stiffness in Pilot Officer Brian Drummond's joints as he sank into one of several canvas chairs in the largest, although small, of the Dispersal building rooms. Squadron Leader Horatio Darling occupied the only other room in the small building that served as his office and the squadron office. This day, Darling as well as half the squadron pilots departed at midday while Green and Yellow Sections were airborne on patrol and engaged in mock combat. The sky was darkening quickly as the six returning pilots let the warmth envelop them to banish the cold and replace it with the mush of fatigue.

"I must say, 'Jackstay,'" said Flight Lieutenant John 'Waggle' Davies, the short, scrappy, determined Welshman, "I think we need to rename your Yank."

"Is that so?"

Brian listened with interest wondering what the Yellow Section leader was going to say. The flight had been a good one. Brian had all 'successes' during the mock engagements with Yellow Section. He suspected he was about to be the target of criticism for leaving the attack to cover his own section leader.

Looking directly at Brian, 'Waggle' Davies nearly growled. "'BAD,' simply does not do him justice."

"I don't know, boss," interjected the formidable, brusque and aggressive Flying Sergeant Miles 'Fog' Johnson, the right wing of Yellow Section, "he is pretty bad."

"Indeed," Davies continued returning his gaze to Brian, "but, I think he lives up to Richthofen's adage about being a hunter. He only had two passes at the banner, and he's already scored more than any of us. I think it would be permissible, since he is young and a newbie to Fighter Command, to change his callsign and call him, 'Hunter.' I haven't seen flying like that . . . well, since I don't know when. What do you think? Should we call him, 'Hunter?'"

"Here. Here," chimed in Davies' wingman.

"Secret ballot," Beamish requested. The assembled pilots held up their right thumb, grasped it with their left hand, as if it was a secret which direction it pointed, and issued forth a resounding raspberry noise. "So, be it. Pilot Officer Brian Drummond, you are hereby ordered to report to the mess at 18:15, sharp, to be properly christened prior to evening meal."

While he did not relish the thought of being drenched with beer, he was proud of his flying and the recognition of his mates. "As you wish, sir."

"With that business dispensed," Beamish began as the lead section leader, "any relevant comments from this afternoon's sortie?"

Davies raised his right index finger indicating his desire to make a comment. "You know, Roger, after our little tiff of a week ago, I think our young Mister Drummond may be on to something. The close quarters of our attack formations may work well for control of the flight, and they may be adequate for Sopwith Camels against Fokker Triplanes. I'm afraid I must side with our colonial. We need all the eyes we can get looking for bandits. If we have just one set of eyes on the leader, then he cannot be looking for Gerry. These attack formations aren't going to work with the speeds of our fighters against the one oh nine."

"We've been over this before, John. We're part of a team, and we have to function within the constraints of the team."

"Somehow we must influence Fighter Command. I think we all can see what works and what does not," added 'Fog' Johnson.

Brian listened, as the newest member of No.609 Squadron, with trepidation for causing the conflict of tactics, but also with excitement that he was being listened to. He was an important element of this group of fighter pilots. Brian liked that. He felt for the first time the power of camaraderie Malcolm had told him about so many times. He felt part of a larger unity.

The discussion among the pilots, even though it was less than half of the full squadron complement of pilots, carried on in earnest until the arguments began to repeat themselves. The level of frustration rose as well as the intolerance of change.

"That's enough, now," interjected Flight Lieutenant Beamish, "we don't need to be bashing one another about this. Unless anyone has something more to contribute, I'd say the debriefing is complete."

"Excuse me," came the delightful voice of Corporal Warren, "Squadron Leader Darling asked me to inform you, after your patrol debriefing, that Pilot Officer Oreland has transferred to Coastal Command effective today."

Pilot Officer Samuel Oreland, the left wing in 'Sparky' Morrow's Red Section, consistently had difficulty with the speeds, the intensity, and the danger of fighter operations. He had not been particularly bashful about his anxiety, concern and reservations about facing aggressive German fighter pilots firing real bullets. He wanted the more sedate mission of Coastal Command. Sam's feeling spilled over so obviously into his flying. He had been the most tentative, the most reluctant of the pilots finding every excuse possible not to engage a target or adversary. Everyone knew he was better suited to a less intense task.

"Poor sod," commented Flying Officer 'Organ' Foxworth, "never did like speed."

The telephone rang. The room fell silent as Flight Lieutenant 'Jackstay' Beamish answered the phone as the acting squadron leader. It would either be a launch command or a standdown order.

Beamish listened for what seemed like the longest time. "Yes, sir." He returned the handset to the cradle. The lack of urgency told them it was the latter command. "All right, lads. It is time to retire to the mess, if you will. Corporal Warren, Dispersal is all yours once again."

"Oh, thank you, sir," she responded feigning excitement.

The noises of grateful men released from a hard day's work mixed with the sounds of the flight kit being passed from bodies to respective pegs. As several of the pilots departed, a young, refined, well-dressed man entered the room.

"Excuse me, gentlemen," the man said with confidence and Brian sensed a sliver of disdain. The accent was distinctly American. Everyone stopped to give the man their attention. "I was told I could find a young American by the name of, Brian Arthur Drummond, here."

"That's it, 'Hunter,'" said 'Mongo' Strickland as a brother acknowledging his siblings discovery as a violator of some silly rule, "you are in trouble now." Brian liked his new moniker. It had a ring to it. He forgot 'Mongo's jab.

"I'm Drummond."

The man looked around the room at each of the pilots staring back at him then at Brian. "May I speak with you in private?"

"Naughty, naughty," 'Mongo' added as he led the other pilots out of the Dispersal building.

"See you at the Mess."

"If you would, sir," added Corporal Warren as she left, "please make sure you turn out the lights and close the door." Brian nodded.

Within a few seconds, they were alone. Brian waited for the man to begin. He extended his right hand. "Good afternoon, Brian. My name is, Arnold Slaughter. I'm the deputy commercial attaché from the embassy in London."

"Nice to meet you," Brian responded with a friendly but slightly suspicious voice as if to say, what do you want with me.

"I'm here on behalf of your parents and the U.S. government," he said with a smile that Brian thought looked like a weasel before it pounced upon a hapless mouse. Brian also knew precisely what was coming next. "It is my duty to inform you that you are in direct violation of Federal law, namely the Neutrality Act of 1937 and 1939. You are directed to surrender at the embassy for repatriation and possible prosecution. You must give up this foolishness."

The tone of the words fueled Brian's anger and resentment of those who did not understand him. For a brief moment, Brian questioned whether this man could actually take him away from his dream. At this moment, he wished he had discussed the possibility with Malcolm, so he would know what to do. In the end, Brian figured this embassy man could not be appreciably different from the police or the border agents in Detroit. "You're wasting your time," he answered harshly and began to walk out of Dispersal.

"Wait a minute," Slaughter said abruptly, "you are not twenty-one years old. You are only eighteen. The Wichita Police have a missing persons report out on you, and your parents want you home. Furthermore, as I said, you have violated the Neutrality Act, and we could easily obtain a federal warrant for your arrest, if you do not come voluntarily."

"This is where I want to be, and this is what I want to do." Brian felt defeated. After all the obstacles he had overcome, he could see the edge of the cliff for the first time.

Arnold Slaughter laughed loudly which was not a good thing to do. "I really don't think you know what you're doing."

The reaction solidified Brian's resolve and made him angry. "I'm flying Spitfires and defending freedom against the Nazis."

Brian's statement sparked another spate of laughter from the diplomat. "Now I know you're crazy. First, flying these contraptions is clearly dangerous. Second, the Germans have not attacked this country. Third, Chancellor Hitler is pursuing legitimate, peaceful objectives."

"You're calling the invasion of Poland a peaceful objective?" Brian shouted.

"It most certainly is. The Germans were provoked by the Poles, and they had to take action to defend their citizens."

"Taking the whole damn country?"

"It may be a bit excessive, but I can assure you, as Ambassador Kennedy has said, Chancellor Hitler is an honorable man who is simply trying to help his people recover from the retribution of Versailles and to protect his legitimate national interests. Ambassador Kennedy has an assurance from the Chancellor that he seeks no other territory in Europe. So, there won't be any war with Germany."

"That's not what Mister Churchill says."

Again, laughter. "And, I'm sure you realize Churchill is a capricious, saber-rattling, war monger who has no appreciation for the subtleties of diplomacy and international relations, so I wouldn't listen to him if I were you."

"My parents taught me to always be polite and courteous, but I am losing my patience and my ability to abide by their teachings. I am an officer in the Royal Air Force. I am doing exactly what I have dreamed of doing. So, I would suggest you leave here before I lose my temper and do something my parents would not be proud of."

"No need to become hostile, Drummond. We are simply trying to help you and keep you from getting hurt. You should remember you have violated Federal law and can be prosecuted."

"Is that a fact? If there isn't going to be a war, how can I get hurt?"

The question stopped the young foreign service man for just a moment. "By flying these grotesque machines," he snapped.

"Now, I really am getting angry." Brian took a step toward him that convinced the man it was time to leave.

"I'll give you a week, and then I will return to retrieve you."

"Over my dead body."

"Anyway we have to, then."

Brian took another step toward him. Without further hesitation, Arnold Slaughter departed. The young RAF fighter pilot stood motionless lost in his incredulous thoughts. He could not believe anyone would want to spoil the realization of his dreams. He also knew his parents were well-intentioned. They cared. They loved him. He just wished they understood his flying and his dream to fly the very aircraft he flew today. Why couldn't they understand and accept his dream? He needed some advice, some help.

"Flight Lieutenant Beamish," Brian said as he walked into the Officer's Mess and waited for eye contact, "I need some help."

"What is it, 'Hunter,' your embassy buddy giving you a hard time?" asked 'Mongo' Strickland. Brian looked frustrated and embarrassed. "Oh, that must be it."

"Let's go outside," Beamish suggested. "What do you have?"

"The guy from the embassy told me he was here to send me back to America and maybe even arrest me if I didn't cooperate."

"So."

"Can he do that?"

"No. You're eighteen and legally able to make your own decisions, plus you are in the RAF. You are actually eighteen, aren't you? You haven't lied on your application, have you?"

"No. I'm eighteen."

"Then, you're OK. I might suggest you contact your benefactor, maybe he can stop this before it goes any further."

"Group Captain Spencer?"

"If that is who he is?" Brian nodded. "Right, then. Now let's get your christening done so we can eat and go to town."

Pilot Officer Brian Drummond was officially anointed with his new callsign. The evening meal passed quickly with a light conversation about flying which was the only common topic, and Brian's visitor that was the hot topic of the moment.

The others left shortly after the meal to enjoy a typical night with the fun loving populace of Edinburgh. Brian remained to deal with his beer soaked uniform after his telephone call.

"Spencer's," came the soft, warm voice of Mary Spencer through the receiver.

"Missus Spencer, this is Brian Drummond. How are you this evening?"

"Lonely, Brian, but other than that quite well I suppose. It is so good to hear your voice. Where are you?"

"At RAF Drem, outside Edinburgh."

"Too bad," she said making Brian wonder about what was on her mind. "What can I do for you?"

"I'm trying to reach Group Captain Spencer. May I speak with him?"

A delicate giggle offered more questions. "You certainly could if he was here, but alas he is not. He rarely is here these days with this damn pretend war around us. You can probably reach him at Bentley Priory."

"Thank you, ma'am."

"Two things before you go, Brian. First, please don't call me Missus Spencer or ma'am. My given name is Mary. I would prefer we were on a first name basis."

"OK."

"Second, next time you are going to be in London, give me a call. Maybe we could have a bite to eat and a night on the town. It is quite lonely around here."

"I'll try."

"Good. I look forward to it. Bye for now," she said and hung up.

Brian placed the receiver back on the cradle. A clear image of Mary Spencer returned to his thoughts. She was certainly an attractive woman much younger than John Spencer and younger by comparison to Gertrude Bainbridge. Brian struggled with the sense of yearning in her voice. He pushed the thoughts back and dialed again.

"Fighter Command. May I help you?" the female voice announced.

"This is Pilot Officer Drummond calling for Group Captain Spencer."

"One moment, please."

Finding John Spencer took several minutes. "What a pleasant surprise, Brian. Good to hear from you. Is everything all right?"

"Yes, sir, I guess."

"Let's have it, then."

Brian provided all the details he could remember about his conversation with the obnoxious diplomat Arnold Slaughter along with the advice offered by his section leader, Flight Lieutenant Roger Beamish. "What can I do?"

"What do you want to do?"

"I want to fly Spitfires for the RAF and beat the Germans."

Again, a chuckle greeted his words. Brian could not avoid the concern about why everyone seemed to laugh at his words.

"Well, you are flying Spitfires as I recall, but I am sorry I cannot order up the Germans. They shall do that part on their own damnable time, I'm afraid."

"You know what I mean, sir. I just want to fly fighters and keep the

Germans from taking Europe."

"Quite admirable, thank God. I shall make a few inquiries tomorrow. Maybe I can keep the bastards off your back. I might add, Brian, you are eighteen and in Great Britain that means you are of legal age. There is nothing they can legally do to you until you return to the United States."

"That's good."

"I would like to add for your benefit, Brian, this young foreign service officer is gravely mistaken about Hitler. Unfortunately, he puts too much faith in Joe Kennedy. There are a few in this country, and I suspect in America, who look upon Kennedy's pontification as near traitorous and in the least quite harmful to peace. But, the bastard is popular with some segments of the population. I just hope not the right parts."

The observations brought sufficient focus to Slaughter's comments as well as all the things he had heard from Malcolm Bainbridge and most vividly his introduction to Winston Churchill himself. Brian knew John Spencer was right.

"Thank you, sir."

"You are quite welcome, Brian. Now, you concentrate on honing the skills that will keep you alive in the coming fight. I will do my best to keep the bastards out of your way. If this guy, Slaughter, or any other sniveller from the embassy, comes back, just ignore them. They have no legal jurisdiction over you. If they try anything fancy, call me immediately."

"Thank you, sir."

"Keep in touch, Brian," John Spencer said as he hung up.

—

Saturday, 9.December.1939
Headquarters, Secret Intelligence Service
No.21 Queen Anne's Gate
Westminster, London, England

Colonel Stewart Menzies, now officially carrying the mantle of 'C' as Director General, occupied a relatively small, austere office. As the chief of MI6, he now sat at the pinnacle of the intelligence community in the very seat held by the first director Captain Sir George Mansfield Smith-Cumming, KCMG CB, for whom the revered moniker of 'C' was preserved. Although politically astute, Stewart chose not to play the games of the political climbers. His reputation rested firmly on a solid, deep foundation of practiced performance within the intelligence community. This cold, dreary Saturday afternoon was no different from all the other extraordinary impromptu sessions instigated by some event or another.

'C's good friend and longtime colleague, Vice Admiral Sir Geoffrey

'Jumper' Pike, Director of Naval Intelligence, was the first to arrive as was usually the case. "Good afternoon, Geoff."

"And good afternoon to you, Stewart. Has Mister Denniston arrived with the hot one?"

"No. I am expecting him anytime now. I am also expecting Winston, and he said he was bringing Bill Stephenson as well. According to Alastair, this one may require some thought."

"Thank God, we have that little box. As most of us felt when first presented with the possibility, thanks to the Poles and Heydrich's Enigma box, we may have the single greatest tool for ending this bloody conflict before it really gets messy."

The buzzer on the interphone rang which 'C' answered. Alastair Denniston, Head of the Government Code and Cipher School, often referred to as the Golf, Cheese and Chess Society in less professional environments, entered the room with a leather satchel chained to his left wrist.

With amenities dispensed, Denniston retrieved a single piece of paper from the case and handed it to 'C.'

MOST SECRET - ULTRA

```
DATE: 29TH NOVEMBER 1939
TO: OKH, OKL, OKM
FROM: OKW
LEADER'S DIRECTIVE NUMBER 4
OPERATION RENE
        OUR LEADER HAS DECIDED TO ATTACK WEST.
OUR OBJECTIVE IS TO SUBDUE FRANCE AS QUICKLY
AS POSSIBLE.  YOU ARE DIRECTED TO PREPARE
COORDINATED PLANS WITH A POSSIBLE EXECUTION
ORDER FOR THE COMING SPRING.  ALL EFFORTS
SHOULD BE DIRECTED TOWARD DEFEATING THE ENEMY
AS WELL AS SECURING CHANNEL PORTS AS QUICKLY AS
POSSIBLE AND PRECLUDE REINFORCEMENT.  INITIAL
OPERATIONS PLANS WILL BE REVIEWED BY THIS
HEADQUARTERS NO LATER THAN 15TH JANUARY 1940.
END
```

MOST SECRET - ULTRA

"We received this message four days ago. You will note Hitler's directive is dated the 29th of November. We broke the sequence three hours ago just before I called you." Denniston paused to give the two intelligence chiefs time to read the message. "We believe the message is genuine although we have certainly asked ourselves the question regarding compromise."

Menzies handed the paper to Pike. 'C' waited for him to finish reading. "What do you think?"

"Is it real?" asked Sir Geoffrey.

"We have no corroborating information, as yet," 'C' responded calmly.

Pike considered the information. "I tend to agree with Denniston and his fellows. If the contents were a plant, there would be no way for the enemy to validate its compromise short of our referencing the specific communication. I believe it is genuine as well, which brings us to an even darker issue. What can we do with the information?"

"I suppose Winston will have a view. The shame of it is, we now have positive proof of Hitler's intentions." Stewart smiled which he rarely did. "I would love to confront that bloody Kennedy with this jewel. Maybe the recognition of Hitler's next move will convince the Americans to join us."

"Maybe, but we cannot use an ULTRA message directly. We must protect Enigma. This note is certainly evidence enough as to the importance of the box," Pike said nodding toward Denniston.

"Quite right." The interphone rang again. The First Lord of the Admiralty Winston Churchill, along with William Stephenson, special envoy to President Roosevelt, joined the small group.

"Welcome to our merry little group, First Lord," said Menzies. "Good to see you again, Bill." Stephenson nodded his recognition not wanting to upstage his friend.

"Thank you, Stewart. Good afternoon, gentlemen. So, 'C' what have we?" asked Churchill.

"In many ways, the message is not good. In a few, it is good." 'C' handed the latest ULTRA message to the First Lord who read it promptly with a scowl on his face and a sunken depth to his eyes. Everyone let the silence linger as a fertile ground for proper contemplation. After several minutes, 'C' continued. "The good is, the productivity of the cypher unit at Bletchley Park is beginning to bear fruit quite well. The group still struggles with the new sequences when the Nazis change them, but their rate of breakdown is improving with each change. Our interception technique is also improving." 'C' paused for a moment. "The bad is, quite obviously, the *Reichkanzler* has apparently decided to invade the Low Countries and France this spring."

The words seemed to slump Churchill's shoulders even more as if an enormous weight had just been heaped upon him. Each man in the room knew why the words had such a noticeable effect on the First Lord. The reality Churchill predicted more than five years ago was coming to regrettable, disturbing fruition virtually in its ugly entirety. No one, least of all Churchill, gained any satisfaction from the knowledge.

"The hard question is, First Lord," began Sir Geoffrey, "what can we do with this information. Is there anything we can do within His Majesty's Government?"

Winston snorted. "After all that has happened, there remains an inordinate number of appeasers and blind peace preservers especially in the Foreign Office. I would love to show this to the President. Convincing the Americans to join us would have far more influence than anything we could do within the government, but alas, as we agreed, we cannot risk compromising our golden egg."

"Quite right," responded Menzies. "Does anyone think we can use this information indirectly, say in a veiled public statement about containing the fighting? Can we alert the Dutch, Belgians and French to the potential invasion?"

"I do not think that would be prudent at any level," answered Admiral Pike. "There are far too many sympathizers in the Dutch government and the French seem to have gone soft."

"I would suggest that assessment might be most unreasonable, 'Jumper,'" interjected Churchill. "The French are no more soft than we are. In fact, I believe they may be slightly more resolved to stop this Teutonic madman. After all, they do not have a channel to dampen the onslaught."

"As you say, First Lord. My apologies, but my recommendation remains the same."

"What do you think, Bill?" asked Winston.

"I'm sure you know how I feel. There is no gain worth even the remotest risk of compromising Enigma. First and foremost, that box and all its products and derivatives must be protected at all costs. About the only actions potentially available to us are: one, tighten our resolve to prepare for what we all fear and now seem to have proof of; two, we must redouble our efforts to find corroborating evidence which would allow us to gradually open up the warning without any connection to this directive; and third, I would suggest a private note from you, Winston, to our friend, referring to further evidence of the gathering storm."

Again, the room fell silent in thought. Admiral Pike spoke first. "I am afraid I agree with Bill. That is about the most we can do. I would only add the caution that we must have sufficient smoke to obscure any connection or even possible connection with any high source in the Nazi government."

"I agree as well," offered Colonel Menzies.

"Mister Denniston, your opinion?" asked Winston Churchill.

"I see nothing inconsistent with the information. I must reinforce Sir Geoffrey's caution regarding protection of Enigma."

"Well, then, gentlemen, it would appear we are concluded."

"If you will permit me," said Denniston, "we finally were able to break down the messages from August and September. The 'excuse' for the invasion of Poland was, indeed, Heydrich's brainchild. An elite SS unit attacked that radio station in Danzig. They arranged for a band of state prisoners to be dressed in Polish uniforms, injected with a lethal dose of Skophedal and riddled with bullets. The bodies were spread around the compound. What's even more morbid, the messages refer to the prisoners as, 'canned goods.'"

"Dreadful," answered Winston. "Do we know any more about the bombing in Munich?"

"Bombing?" Stephenson asked.

'C' shook his head. "Not much." He looked to Bill. "On the evening of 8th of last month, a bomb exploded near the rostrum at the *Bergerbraukeller*. The detonation missed Hitler by 13 minutes."

"I'll be damned," said Stephenson.

"The Gestapo arrested a man by the name of Elser, Georg Elser, reported to be a cabinet maker and musician. For reasons we do not know, Hitler cut short his usual two, two and a half, hour celebratory speech. You will recall it was on the same day in 1923 at the very same beer hall that the Nazi *putsch* began, for which Hitler and his cronies went to Landsberg prison."

"I'll be damned," repeated Stephenson.

"Do you think it was a bona fide assassination attempt?" asked Winston.

"Hard to say, actually. The Gestapo is not particularly forthcoming with accurate information. From our limited sources, this fellow Elser is being held incommunicado and in isolation from everything. The press reported eight killed in the blast and many more injured. There have been no legal charges, although there have been a few press assertions that this fellow was an agent of the British government. Our guess is the Nazis will use their martial law authority to keep this very quiet. I cannot imagine a public trial. They do not want anyone to think a German citizen might be displeased with *Herr* Hitler and his beneficent regime. I suspect he will be summarily executed in the dark of the night someday. . . if he hasn't already."

"Like the Night of the Long Knives," added Churchill.

"Indeed. Yes. So, our guess is, yes, it was a bona fide assassination attempt."

"We know there are others."

"Yes. However, our problem is getting to them without touching nationalistic sensitivities or compromising those who are dissatisfied. If we had moved prior to the Polish invasion, we might have had a chance."

"So, we watch, listen and wait," Winston said.

"I'm afraid so."

The room remained silent for several minutes as each man considered the implications of the information. Each of them suspected a manufactured reason although none of them had guessed anything as grotesque as the reality.

"This should give us a good idea of who we are dealing with," said 'C' finally.

"Most assuredly," answered 'Jumper.'

None of the men wanted or needed to discuss the additional information. The group promptly dispersed after proper *adieux*.

Chapter 16

> Success is relative.
> It is what we can make of the mess
> we have made of things
> -- T.S. Eliot

Monday, 11.December.1939
RAF Drem
Drem, Lothian, Scotland

NEWLY commissioned Pilot Officer Jonathan Andrew Xavier Kensington, otherwise known to his colleagues as 'Harness,' arrived at the airfield that served as home for No.609 Squadron. Having graduated from the advanced training course as his friend Brian Drummond had done nearly two months earlier, Jonathan spent the weekend with his congratulatory and proud family at their home, Carlingon Castle, in the evergreen wooded hills west of Newcastle-upon-Tyne. There was absolutely no question of his excitement, still barely contained, over winning the assignment to the same Spitfire squadron as his best friend. For once, he was immensely grateful for his father's influence and thankful to the Lord for opening a fortuitous seat in Brian's squadron.

The short train ride from Newcastle to Edinburgh, and then back out to Drem on the local train enabled him to check into the squadron shortly after noon. One lone Spitfire, probably not flyable or maybe his aircraft since he was the replacement pilot, occupied the flight line signifying the fact the entire squadron was airborne on patrol. He eagerly awaited the reunion with Brian. The telephone conversation with his friend when he received his orders conveyed his excitement, but did not lessen his anticipation of their meeting.

The administrative formalities were completed in their entirety before the squadron returned from patrol. With his belongings properly stowed in his small room closet, Jonathan checked the nameplates to see where Brian's room was located – three down from his.

Jonathan walked to the flight line looking into the vacant repair hangar. This had to be a good squadron, he told himself. All the aircraft, save one, were flying. That fact spoke well for the maintenance crews. About the time he arrived at the Dispersal building, Jonathan heard the distinctive purr of several Merlin engines at high power. Over the treeline on the small rise to the east came the first three Spitfires in a close left echelon formation at high speed and very low altitude. Just as the leader broke hard to the right followed at timed intervals by his two wingmen, the next flight of three Spitfires repeated the maneuver. Many of the ground crew cheered at the glorious sound and majesty of the squadron's demonstrative return from a successful patrol. Even Jonathan wanted to cheer although he was satisfied with a broad smile as he watched the display before him.

Jonathan Kensington enjoyed every moment of the squadron recovery. Each of the eleven Spitfire fighters completed the landing circuit individually. Like a line of graceful and deadly wolves, the aircraft followed the path of their leader swinging their noses around to be ready for the next launch. Each of the pilots dismounted, exchanged words with their crew chiefs, and began walking back toward Jonathan. He spotted Brian as he extracted himself from the sixth fighter in the row. Brian did not recognize Jonathan until he was more than halfway back to Dispersal. A broad smile filled Brian's face. His shoulders straightened and his pace quickened.

"Hellava great day, this is," Brian shouted from seven yards away causing most of the pilots to look back at him.

"You must be, Pilot Officer Jonathan Kensington," said the squadron leader as the first person to reach him.

"Yes, sir."

"Good to have you with us, 'Harness.'" Jonathan was somewhat surprised that the squadron leader knew his callsign already. "I am Squadron Leader Darling, callsign 'Spike.'"

"A pleasure, sir."

"'Hunter' has been a good advertiser for you."

"'Hunter,' sir?" Jonathan asked not recognizing the name.

"Ah, yes, well, you would have known him as, 'BAD.' We've changed his callsign to, 'Hunter.'"

Jonathan understood immediately but wondered why they changed Brian's professional name. He would know in time. Brian Drummond and Jonathan Kensington shook hands and embraced as brothers. Introductions were completed including Jonathan's assignment to Flight Lieutenant Robert 'Sparky' Morrow's Red Section – C Flight. His aircraft would be the PR-K – 'King' bird. Jonathan listened with interest to the events of the unsuccessful patrol to find a small flight of intruders among the layers of clouds.

Brian remained behind after the evening meal and usual squadron welcoming ritual to talk with Jonathan as he cleaned up. He waited for Jonathan to finish his shower. The pinup girls adorning the walls added the only richness to the austere décor of the pilot's residence.

"It's great to have you here, Jonathan," Brian said as Jonathan finished drying off and began to dress.

"I wasn't sure I was going to make it, but I am glad I tried. Your impression of the squadron certainly convinced me this was where I needed to be. Besides, like you, I would rather be flying Spitfires than Gladiators or anything else for that matter."

"Yes, indeedie. How did you get the change?" Brian asked knowing Jonathan must have pulled some strings as he had done.

"Just about the same way you did. My father is personal friends with several MP's."

"Anyway that works, I would say."

"Agreed."

"You must have gone home upon completion of advanced."

"I am not crazy like you. Yes, I did go home. In fact, didn't Squadron Leader Darling say we should get this coming weekend off?"

"Yeah."

"Why don't you come home with me and meet my family?"

"I'd love to, Jonathan, but I have already called Anne. We're going to get together in London."

"That is a hell of a long way to go for a night in the sack."

They both laughed. "Yeah, well, she's worth it."

"So you say. Then, maybe you can spend the Christmas holiday with us."

"Sure. Sounds just right."

A moment of silence helped Jonathan change the subject. "Do you really think this is the right place to be?"

"What do you mean?"

"Everyone seems to think Hitler will try to come across the Channel although all the action seems to be in the North."

"I'm no expert by any stretch of the imagination, but some of the guys think he might try the North since the defenses are thinner up here. When I met Churchill, he thought the Nawzees," Brian tried to replicate Churchill's slurring of the word, "would try to take all of Europe, especially France, since that was where the Treaty of Versailles was signed. If that happens, then the Channel makes the most sense."

"How does Hitler know our defenses are thin in the North?"

"Hell, they fly reconnaissance flights virtually every day. We've caught a few as you know, but most get through because of the weather, and they fly at very high altitudes. Only the Spits can get up that high, but these two blade airscrews just don't have the push above 25,000 feet. Plus, it doesn't take a genius to notice we have one squadron of Spitfires here, and we have four squadrons, two Spits and two Hurcs, at Hornchurch, for example."

Jonathan sat down on his bed and looked into Brian's eyes. "Shouldn't we be in a One One Group squadron if the action's going to be in the South?"

"Maybe so, however, Darling feels certain we will move south, if Hitler attacks France."

Kensington paused for more thought. Everything Brian said made sense to him. He trusted Brian's instincts although he knew there was no way for him to have any definitive knowledge of what was going to happen. "That should do it, then, don't you think?"

"I'd say."

"Let's go get a pint as long as none gets spilled on me." Both men laughed as they left for the Mess.

Wednesday, 13.December.1939
The Admiralty
Whitehall, London, England

"Good evening, Sir Dudley, Sir Geoffrey," said the First Lord of the Admiralty.

"Good evening, sir," responded Admiral Pound, First Sea Lord, for both senior naval officers. "We have considerable information for tonight's briefing."

"Very well, then. Let's get to it."

"Sir Geoffrey, if you would be so kind," said Pound.

"First, if you will recall the sinking of the armed merchant *Rawalpindi* on the 23rd of November," Pike paused to receive a nod of confirmation from Winston, "we have determined from a variety of sources that it was not the pocket battleship *Deutschland* that was involved, as originally reported, but the battlecruisers *Scharnhorst* and *Gneisenau*. Not that it alters the result; the information enables us to establish a clearer pattern for the German capital ships. We are attempting to use the patterns to anticipate future actions."

"Excellent."

"Second, before we address the primary news, I might add the Finns are holding their own rather well against enormously superior odds. All available reports show us an exemplary use of terrain and weather to advantage against the beleaguered Soviets. The losses sustained by the attackers have been enormous although the Soviets have shown no indications of reducing the pressure on the Finns."

"Good for Mannerheim," said Winston. "Just shows you what the home pitch can do for an inferior force. I hope the lessons are not lost on us."

"Quite," answered Sir Dudley. "Most important is the news from the

South Atlantic. Sir Geoffrey, if you would continue."

"At 06:20 hours this morning, Commodore Harwood, Henry Harwood, in command of Force G aboard *Ajax* along with the cruisers *Exeter* and *Achilles* discovered the pocket battleship *Graf Spee* off the River Plate Estuary. A ferocious surface battle ensued for 90 minutes until the German decided he had enough and broke off the engagement. The *Spee* made for the neutral harbor of Montevideo."

"Damage?"

"All three of His Majesty's ships suffered significant damage, however they remain operational. A blockade has been established to prevent the *Graf Spee* from escaping to the open sea and all available South Atlantic Fleet ships have been requested to join the blockade. All three cruisers need repairs but will not withdraw until they are satisfactorily relieved on station."

"Dear God Almighty, we finally have ourselves a bona fide naval victory," Winston said as he shook both fists. "It is about bloody time. Sir Dudley, please cable our sincerest and most rejoiceful congratulations to Commodore Harwood and his men. Job well done."

"The battle may not be over," added Sir Dudley in a soft voice as if he might embarrass his minister. "The *Graf Spee* has been damaged, but she is far from disabled."

"What is her likely move?"

"They will undoubtedly assess her damage, seaworthiness and battle capability, as well as seek guidance from Berlin."

Winston immediately thought of ULTRA and prayed they may have finally broken the difficult naval codes. He looked to Pike who only discretely shook his head. "Will she come out fighting?"

"Her captain, Landsdorff I believe, knows he has only a few days at most to attempt a break-out. After that, he must recognize he would face an overwhelming naval force," said Sir Dudley. "The next few days are crucial."

"We must make every resource available."

"We have already done so, First Lord. The cruiser *Cumberland* has terminated her refit and departed the Falklands to join the blockade. She is a little more than a day's steaming away. The cruiser *Renown* and aircraft carrier *Ark Royal* from Force K have put into Rio de Janeiro for prompt refueling and should be on station within two days. The cruisers *Dorsetshire* and *Shropshire* from Force H departed Capetown several hours ago and are three to four days out. As I said, First Lord, the next few days are absolutely crucial."

"Brilliant. Simply Brilliant. Good fortunate has brought us a prize worthy of the chase. We cannot allow her to escape our grasp."

"The entire Royal Navy appreciates the importance, and we shall prove ourselves up to the task."

"Excellent. Absolutely brilliant. I must inform the Prime Minister. Please keep me informed of every dispatch as they arrive. I intend to remain here until this action has been concluded."

"As you wish, First Lord."

Churchill returned his thoughts about ULTRA and wondered if it was or could be any help in achieving final victory over the illusive and highly successful German raider. He remembered the difficulty GCCS had breaking the naval versions of the Enigma codes, primarily because of the various internal code words and jargon used by the German naval communications personnel. He wanted to ask Sir Geoffrey but any discussion could easily exceed the clearance authority even for the First Sea Lord who had only limited access. He would try to find a private moment with Admiral Pike who remained one of the very few people allowed full access to ULTRA information. He decided to take a different tack.

"Has the Foreign Office been informed? And, have they initiated diplomatic pressure upon the Uruguayan government to expel the German combatant?" asked Winston.

"Not yet," responded Sir Dudley. "We wanted to brief you first. We shall make the information available to them promptly and seek their assistance."

"It would appear the forces should be sufficient to deal with the German raider. Do you agree?"

"Yes, First Lord. The *Cumberland* was nearly complete with her refit. Our reports indicate she is fully fueled and armed although she is short other supplies, none of which should affect her battle readiness. With *Renown* and *Ark Royal*, the noose should be complete."

"As I said, keep me informed and if there is anything I can do, by all means, let me know immediately. Good luck, Sir Dudley."

"Thank you, sir."

Winston thought quickly to fabricate a reason for 'Jumper' Pike to return for a discussion regarding ULTRA. "Sir Geoffrey, if you would be so kind, I would like a thorough briefing on the specifications for the *Graf Spee* in about half an hour," he said as calmly and professionally as he could, given the excitement of the hunt.

"It shall be done," responded Admiral Pike with a knowing glance.

Both senior officers left the First Lord's office to carry out their duties. He immediately called No.10 Downing Street, asked for the Prime Minister, and provided a brief description of the day's action in the South Atlantic. He was careful not to mention His Majesty's ships involved or the reinforcements dispatched to assist Commodore Harwood in the prosecution of his mission. Winston could readily sense the lack of excitement and enthusiasm in the Prime Minister's voice. The reaction was no different from the report of a reconnaissance bomber overhead London.

He felt the excitement of the first real taste of success since the war began three months earlier. He wanted to share his excitement with the world, but knew it was premature. The next call he placed went to Chartwell. "Mister Smithfield, I should like to talk with Missus Churchill, if she is available."

"Certainly, sir. One moment, please." The task did not take long.

"Yes, Winnie," the familiar voice came to him.

"Clemmie, we have a major naval battle begun in the South Atlantic. I shall need to remain at the Admiralty until it is concluded. I trust you will appreciate the need."

"Is it good?"

"Yes, Clemmie, it appears to be very good. I want to remain close to the action. If you need me, do not hesitate to call," he said although they both knew the only call that would come through would be a life-or-death emergency. Clementine Churchill respected her husband's sense of duty as she had suffered all the sacrifices necessary with him.

"Good luck, Winnie. Our prayers are with you and the Navy. May the good Lord bring us victory."

Winston's eyes welled up with pride in his wife. She understood the world and him, and for that he was immensely thankful.

"Thank you, dearest. I shall try to call as this situation progresses. Until then, be comfortable."

"Not to worry. We shall be here when you are done."

―

Saturday, 16.December.1939
Westminster, London, England

"This was a bad idea," Brian heard Jonathan Kensington say as he was standing under the eaves of Kings Cross Station watching the unusually heavy rain flood the street and add greater chill to the air. Occasionally, even a few snowflakes could be seen. The train ride from Edinburgh had taken the greater portion of the day. The cold evening air did not soothe their stiff joints and muscles.

"Now, now, Jonathan. We'll have great fun once we cross town."

"Maybe for you, but there is no guarantee Anne will have a friend to my liking."

"Ah, you're just a worry wart. She will do the task brilliantly, and we are going to have a smashing good time."

"Aye, mates," an old, stubbled faced, slumped over, man shouted to the two RAF pilots above the din of the rain. "You lads keep up the good work and keep the Krauts from over our 'eads."

"We'll certainly do our best, sir."

"Aye, certain you will, me is," the man said as he squeezed Jonathan's elbow and shuffled out into the rain.

The two men watched the old man move away. Other citizens, male and female, young and old, nodded, saluted, tipped their hats, or otherwise acknowledged the presence of the uniforms that would soon defend them against an intimidating enemy.

"This recognition never ceases to amaze me," Brian said.

"I know what you mean. It does make you feel good about what we are doing, doesn't it?"

"Absolutely."

"Enough of this drivel. Are we going to wait here for the heavens to close, or are we going to get on with the night's festivities?" Jonathan said, and then darted from under the eaves along Euston Road heading toward the stairway for the Underground station. Brian followed his lead.

The Piccadilly Line tube train carried Brian and Jonathan south from Kings Cross to the Knightsbridge Underground Station. The distinctive façade of the famous Harrods department store greeted them along with the diminished rain of the chilly December evening. The short walk to Beauchamp Place positioned them at the foot of the steps leading to Anne Booth's townhouse. Without words, Brian looked at Jonathan, took a deep breath and ascended the steps. Brian's heart rate stepped up a notch as he rapped the brass doorknocker.

The door opened. Within an instant of recognition, Anne wrapped herself around Brian. A brief but passionate kiss completed the greeting. "You have made me wait too long for this moment, Brian Drummond," Anne said with another kiss. "Good to see you again, Jonathan."

"A pleasure, Miss Booth."

"Come in. Come in. Linda Mason is eager to meet you," Anne said to Jonathan with her arm around Brian's waist and looking over her shoulder.

"Great."

Brian detected a touch of sarcasm in his friend's response. He looked back at his friend giving him an expression to encourage him to loosen up and enjoy the moment. Jonathan returned an expression of resignation regarding his blind date.

The familiar living room brought its own warmth. An attractive, smiling, blond woman stood as they entered the room. "Linda, this is my man, Brian Drummond, and this is his friend, Jonathan Kensington."

"A delight to meet you, Brian," Linda said with a smooth, clear voice as she extended her hand to Brian. She extended her hand to Jonathan accepting his with both her hands. "I am truly excited to finally meet you, Jonathan."

"Likewise, I am sure."

With their expected arrival time, Anne prepared a light meal for the four of them. The conversation was airy and humorous about subjects unrelated to the war and the looming danger. London's tabloid gossip dominated the talk until they retired to the living room.

"So, what do we owe the honor of the visit of two RAF pilots from Scotland?" asked Anne as they sat with drinks in hand.

Brian stood, picked up a letter opener and tapped his glass as if he were gaining the attention of an audience. "We are celebrating Jonathan Kensington's completion of the advanced course, his promotion and commissioning as a pilot officer, and," Brian emphasized, "his joining of the best fighter squadron in all of Great Britain."

The proper congratulatory words from the two women seemed to embarrass Jonathan as he nodded and swirled his drink.

"Is that unusual or lucky?" Linda Mason asked genuinely curious.

"Actually, no," responded Jonathan. "I had some help, as Brian did, to switch from Gladiators to Spitfires. In my case, my father and his friends helped me. It was fortuitous that Six Zero Nine Squadron happened to become short one pilot when I came up for assignment."

"Brilliant, simply brilliant," said Anne.

"That Spitfire is such a beautiful aeroplane," offered Linda. "It is such a sleek machine."

Both young men reflected a sliver of surprise at the frank words while both women smiled brightly.

"Yes, I suppose you are correct."

"Absolutely. That's why we fly them."

"Is it really that much fun to fly them?" asked Linda.

"It's the next best thing to sex," Brian said with emphasis.

"Do you agree, Jonathan?" asked Linda.

"Without question."

"Well, then, Anne, I suppose we have our work cut out for us if we are to retain number one status."

"Exactly."

"To give you an idea how important it is, Brian decided . . . ."

"Jonathan, come on now. You don't need to go into this."

"Oh yes," Linda said with an air of excitement and revelation to come. "We want to know."

"Brian decided to pass up coming to London to see Anne when he completed the advanced course, in favor of proceeding directly to RAF Drem to join the squadron. That's how important flying Spitfires is."

"Oh my."

Everyone looked to Anne. "Now, now, Brian has explained. We've been together twice since then. He has certainly made it up to me," Anne said squirming a little as though she had been suddenly aroused. The laughter acknowledged the message.

"But, isn't it dangerous?"

"Not really," answered Jonathan. "It is more dangerous driving a motorcar down the carriageway."

"That can't be true," came Linda's incredulous challenge.

"We think it is," added Brian.

"Have you been in combat?" asked Linda, back to her curiosity.

"No, not yet. We've been on patrol, and we've had a couple of scrambles, but I haven't fired my guns at the enemy, yet. Jonathan just joined the squadron this week."

All eyes turned to Linda anticipating the next question. She thought about Brian's answer to form her question. "Is there really going to be a war?" she asked.

"There is a war," Jonathan said strongly. "The Navy has been rather busy. The bloody Germans sank the *Royal Oak* in the middle of Scapa Flow, for God's sake. The Air Force constantly has to deal with German reconnaissance aircraft overflying the United Kingdom. There is no reason the Germans would be over our cities if they did not intend to use the information they are gathering. They are preparing their bombing target list. I should say the war is going to get much closer in the not too distant future."

Silence provided the collective impact to Pilot Officer Jonathan Kensington's words. Brian knew he was precisely correct. All the pilots did. All the King's military men and women knew what was most likely going to happen although none of them wanted it to happen. Each of them sipped their drinks as they contemplated the developing reality surrounding Jonathan's description.

Linda Mason broached the silence of contemplation. "Are you serious?"

"Yes."

"They are going to bomb London?"

Brian joined the discussion. "Many believe it is only a matter of time. We are preparing for whatever might happen. London is an obvious target."

"That cannot be true."

"Maybe not, but we are training for what we have been told to expect, the full weight of the *Luftwaffe* over Britain."

"Oh my God. What will we do?"

"We will persevere and we will prevail," interjected Anne for the first time adding a profoundly British twist to the words.

"Quite," responded Jonathan.

"We've got some of the best fighter pilots in the world and the best fighters. We will do what we have to do. We will defend Britain."

"I think it may be time to change the subject. This topic is too depressing no matter how close, or how real it may be," Anne said.

"What do you suggest?" asked Linda.

"Actually, I think it is time to make love, not war," offered Anne.

"Here, here," Brian laughed. He noticed Jonathan's mood change. Brian could only surmise Jonathan's discomfort grew from the suggestion of intimacy with Linda Mason, whom he met just a few hours earlier. He probably did not want to face the next step between them after such a short time interval.

"If there are no objections, we'll leave you two to get more acquainted without an audience," suggested Anne.

"You two go ahead. Jonathan and I will be quite all right, I should think."

Anne took Brian's hand and led him to her bedroom that had now become familiar, a place he felt very comfortable in. She closed the door behind them and promptly wrapped herself around him, again, kissing him deeply and passionately. Brian responded with equal swiftness. Satisfied with the result, Anne withdrew as if a dark thought suddenly came to her. She turned to face him now several yards across the room. "Do you really think they are going to bomb London?"

"That's a hellava question at a time like this."

"Yes, but do you?"

"I don't sit at Hitler's staff table, Anne. I don't know. All I know is what we are preparing for and what we hear from people who know a lot more than I do. It sure does appear we are headed that way."

"You and your mates have to protect the city, Brian. Innocent people will get hurt."

"Anne, Fighter Command is not large, but it is good and determined. We are getting better every day."

"But, you are so young to have such a horrendous job."

"Maybe, but I am by no means the youngest. I've heard of some pilots in other squadrons that are two years younger than me. I am the youngest in my squadron. Nonetheless, we'll do the best we can and that is all we can do."

"It is so dreadful. Why must we have war? I can remember my parents talking about the horrors of the Great War and especially when the Zeppelins bombed London."

Brian could only imagine what her words meant. He felt heavy and in some ways inadequate to reassure Anne. He also felt helpless, not knowing what to do next. He stood there motionless and speechless waiting for Anne to take the next step.

"We need to change the subject," Anne finally said as she moved toward Brian with a gradually spreading smile. "We have much better things to do than lament about this foolish war." Anne started to undress him as he reciprocated. "No. Just let me take care of everything. Simply do what I tell you."

Her gentle motion and expert hands brought him to full readiness by the time he was completely naked before her. She took him in her soft, warm hands stroking him ever so slowly until he reached for her.

"No," she repeated.

Brian stopped and lowered his arms. The urge to take her grew with each stroke. He wanted to feel her skin, the soft, smooth curves of her body. Her touch was absolutely delightful, and yet it also brought a building tension and tightness.

Anne directed him to the bed pulling the thick *duvet* and top sheet back into an accordion fold at the foot of the bed. With her hands, she indicated she wanted him to lie on his back in the middle of the bed. From the nightstand, Anne removed several long, colored, silk scarves and tied each extremity leaving Brian spread-eagled. Without taking her eyes from his, Anne deliberately stepped gracefully through her disrobing to tease, titillate and tantalize Brian in his restrained state. The exhibition produced the desired effect.

Anne ascended the bed like a big cat stalking her evening meal. She touched him in numerous key spots just to make sure he could feel her presence. Ever so softly, she swung her breasts allowing her erect nipples to drag across various parts of his body finally letting him take to her proffered nipple. He passed his passion through their only point of contact. Anne purred with contentment at the attention and the noticeably enhanced expertise taught to him over their few significant events of intimacy. As she withdrew from him, Brian pulled against the restraints wanting more as if that was all he was going to get. Anne had a different idea.

Standing fully erect and straddling his torso, Anne lowered herself until the heat and moistness of her could just barely be felt at the point of contact. As he raised his pelvis to join with her, Anne withdrew. Again, she tempted him with identical results.

"You can't keep doing this, you're driving me crazy."

"That is the whole point, my lovely young buck. That is the whole point."

This time Brian remained still. Anne, satisfied she was in full control of their activity, took him in partially, gripping and pulling him to allow him to feel the strength of her muscles. Her slow, measured movements elicited a thrust from Brian to gain a deeper union. Anne immediately withdrew from him.

"Now, now, my eager lover," Anne said playfully castigating Brian.

"My God, woman, I think my dick is going to explode."

"Oh, relax, my sweet, no harm will come to you, only pleasure."

Anne let him suckle at her breast before returning to her playful activity. Brian moaned and groaned as she gradually took more and more of him until they were fully united in the most intimate way. She paused to rest, beads of sweat descending down her body, dripping from the lowest point of their journey. Brian felt the drops tingle like strange slivers of ice. Anne rested her chest upon his with her face nestled against his neck. She kept him stimulated with the rhythmic constriction of her muscles. The sensations drove him to fight his restraints. He wanted to embrace her, to feel the wet, hot smoothness of her body, to stroke her hair, but he could not overcome his bindings. She allowed him to move against her, to take what he could of her.

"It is time to give yourself to me," Anne said as she rose on her hands and feet. With precision, she moved upon him as if she were milking a cow's udder. As she felt him harden toward his climax, her pace quickened undoubtedly knowing the end was near.

Brian screamed as his body shook and every nerve ending flickered between intense pleasure and excruciating pain. He had never felt sensations that had taken him beyond his control. The closeness drew them even closer to one another, and he liked the feelings.

Anne released him from bondage. His arms immediately engulfed her. They lay motionless, speechless and joined while their heavy breathing gradually subsided.

As was the strength of his youth, recovery came quickly to Brian. The playfulness born in the feelings they had for one another enabled them to enjoy each other with ease, laughter and pleasure. He repaid his debt several times over that evening as they enjoyed the pleasures of their passion until sheer exhaustion and near contentment overcame them both.

Sunday, 17.December.1939
The Admiralty
Whitehall, London, England

The cold, dark day gave the First Lord of the Admiralty the first respite from the intensity and action of the Battle of the Rio de la Plata. He spent the better part of nearly four days buried in the Admiralty War Room trying not to interfere with the professional officers performing their duties but reviewing every dispatch from the South Atlantic. ULTRA yielded nothing other than some routine reference in several Air Force messages to the stalemate off shore Uruguay. Diplomatic pressure on the neutral nation produced progress but not results. Winston finally decided to return to his office. The denuded trees of St. James Park looked as cold as they were.

The urge for some fresh air pushed Winston to raise the window. The rush of biting, crisp, cold air cascaded in. He leaned out and inhaled deeply. There was only the faint smell of the city, but the cold air felt good in his lungs. He coughed several times as his lungs protested the shock. The contrast brought by the cold exterior refreshed and invigorated the First Lord who closed the window and returned to the papers on his large desk.

Winston had not seen his wife, Clementine, or his children in more than half a week. He wanted to enjoy the refuge of Chartwell and the comfort of his family, but the grand prize of the most recent naval combat kept him at the Admiralty. He felt he simply could not leave his post with such an important victory so close at hand. He tried to read through the more routine and mundane communications associated with his ministerial position, but found his mind returning to the War Room.

The dispatches from Commodore Harwood gave all of them a taste of the situation. HMS *Ajax* patrolled back and forth just outside the three-mile territorial limit like an impatient wolf waiting for the precise moment to pounce. They could easily observe movement of men above deck on the *Graf Spee*. They were within easy gun range, in fact, main battery shots from the six-inch guns of *Ajax* or the eight-inch guns of *Exeter* and *Cumberland* would be nearly point blank. They all knew the massed guns of Force G could administer the *coup d'grâce* to the German pocket battleship and highly successful surface raider in less than a minute. Seeing the German capital ship so close in the Montevideo harbor had to be frustrating for Harwood and his men, but Winston sensed the end was near.

A knock at the door brought a welcome relief to his administrative agony. "Enter," Winston answered.

Admiral Pound entered the room with a broad smile of his face. "This affair is over."

"What happened?"

"She steamed slowly out of the harbor. Harwood deployed his force for battle, and she blew herself up. Scuttled!"

"Dare say!"

"My response precisely, First Lord. They apparently set several charges in her main magazines and let her blow. She sank immediately, in seconds the report stated."

"What of her crew?"

"As near we can tell, the entire crew disembarked. A small scuttling party was able to escape prior to the explosion."

"Her captain?"

"We received word a short time ago from the Foreign Office . . . he appears to have committed suicide. Single pistol shot to the head, so the report stated."

"How typically German. Terrible waste."

"Indeed."

Winston Churchill reviewed the events in his head as he considered several necessary actions. He wanted to call Clementine but that would have to wait. "If you would be so kind, please send the appropriate congratulatory messages for both of us, the Navy and indeed the entire, grateful nation."

"Absolutely," responded the First Sea Lord.

"Also, would you prepare the proper recommendations for military recognition of this grand achievement. I will inform the Prime Minister and the War Cabinet. Well done, Sir Dudley. Simply Brilliant. Well done."

"Thank you, sir. I will convey your sentiments to Commodore Harwood."

Churchill made all his official calls. The Prime Minister, in a gracious and elegant manner, requested that the First Lord notify the King on behalf of His Majesty's Government. Winston appreciated the recognition Chamberlain gave him and thoroughly enjoyed the short conversation with King George VI who was equally appreciative and thankful.

The victory over the *Graf Spee* represented the first major victory over the Germans since the war began. It was truly a moment to be savored. After the loss of the *Royal Oak* to a U-boat while she was at anchor in the harbor of Scapa Flow two months earlier, this victory was made all the sweeter for the First Lord of the Admiralty.

Chapter 17

> If men knew all that women think,
> they'd be twenty times more daring.
> -- Alphonse Karr

Saturday, 23.December.1939
Waverly Station
Edinburgh, Lothian, Scotland

THE fact that the rail station was nestled in the valley between fairly high hills did little to shield the two pilots from the bitter wind chill of the winter day. The overcast kept the sun from providing any warmth. The walk to the station house convinced them further delay would not be beneficial.

The waiting room of the station house was virtually deserted. Only an elderly couple waited with them. "Tell me, again, why we aren't going south to see Anne and Linda?" Brian asked as they closed the door finally gaining relief from the cold.

"This is a family time of the year. You have no family in Britain, so you are going to spend Christmas with my family. You can contain your lust for a few days."

"Why? You had just as good of a time as I did. You want Linda and you know it. You are just too much of a sod to admit it."

"Now that is simply not true. I would very much like to enjoy the pleasures of Linda's company, but this is a family time. Anyway, we owe you a proper British Christmas since you are a volunteer serving in defense of freedom and this country." The billows of steam from the locomotive announced its presence despite the auditory insulation of the station house. "Here's our train."

"I suppose it's Newcastle instead of London for us, then. So, let's get to it," Brian suggested with a smile and a slight chuckle. Although he jabbed at Jonathan a little, he was actually excited and anticipatory about the coming holiday weekend.

"Yes, indeed. I know you are going to have a delightful time. My family is looking forward to meeting you."

"Oh, great . . . expectations and disappointments," Brian joked as they boarded the train.

The empty compartment allowed the two men the freedom of unconstrained conversation. The journey was filled with a running narration about the scenery, history and culture of the land passing by them, interspersed with male reminiscence about the previous weekend. Even to the two newest pilots of No.609 Squadron, the amount of time allowed for rest and relaxation seemed inordinately high for fighter pilots in the middle of what was supposed to be a war. They appreciated it nonetheless and decided to take full advantage of the time since most of the older officers advised the generous allowance would not likely last once the real war started. Who were they to argue?

The train ride south to Newcastle took just over an hour. A large, black, chauffeur driven Bentley waited for them. The driver was a middle aged man who had been in the long term employ of the Kensington family. Brian had suspected Jonathan came from a wealthy family. The limousine offered the first bona fide evidence his suspicion was correct. Jonathan continued his description of places, people and history as they moved west out of the city. Even with no leaves on the trees and the gray, windy sky, green grass covered the hills, fields and gardens. Brian enjoyed the large swaths of green in contrast to the shades of brown in Kansas.

As the car moved slowly down a narrow country lane, a large stone wall came into view. It seemed to extend east-west as far as he could see.

"It's Hadrian's Wall," Jonathan offered.

"It certainly is an impressive wall. It must have taken a lot of work to get such a wall in place."

Jonathan chuckled at Brian's innocence. "This wall was built nearly two millennia ago by the Roman Legions occupying this country under the Emperor Hadrian. The wall extends virtually the full breadth of England. This was the Northern most reach of the Roman Empire."

"My golly. I don't think I've ever seen anything that old in my life."

"There are other monuments much older than this in Great Britain alone," Jonathan said as they passed through a breach in the wall with apparently much newer stone pillars at either side of the roadway.

"Like what?" Brian asked with genuine curiosity.

"Stonehenge, for one."

"What's that?"

Jonathan chuckled again. "You are quite the innocent, Brian. It is a grouping of massive stones forming an astronomical circle built by the Druids many more thousands of years older than this wall."

The history around him fascinated his young mind. It also made him feel quite small. Feelings of years, deep roots, formidable foundations and strength of character came to Brian in the stones of Hadrian's Wall with its surrounding green grasses and conifer trees among the leaf bare shrubbery and trees.

Jonathan's narration continued although the words slipped by Brian's awareness as comparisons to the relatively brief history of his own country and the state of his birth occupied his thoughts. It did indeed make him feel small, but it also reinforced his commitment to defend this land against those intent upon harm and subjugation despite the protestations of Arnold Slaughter, that weasel of an embassy official.

The first indication they were nearing their destination was a small sign beside the narrow roadway that simply said, Carlingon Castle. The sign brought Brian back to Jonathan's words that had nothing to do with his home. The car turned off the paved road onto an even narrower gravel path barely wide enough for the large limousine. The path weaved through a lush, deep, conifer forest. The variety and richness of the vegetation even in the cold of winter impressed Brian. The earthy, aromatic smells of the forest filled the interior of the car even though all the windows were closed against the cold.

"So, where is your home, Jonathan?"

"It's another few miles, yet."

Brian looked directly into Jonathan's eyes. "Why didn't you tell me your family was wealthy, or important, or something?"

"First, I didn't think it was important to our friendship. Secondly, it is not the proper thing to do, and thirdly, although I am appreciative of what my family's wealth provides, I want to set my own course, not the one my father wants."

Brian recognized the growing familiarity of Jonathan's quiet rebellion, but chose to steer away from the topic. "Maybe, but this is all a bit of a shock to me. I've never been in a place so beautiful and apparently so big."

"I have never really thought of it that way. I just think of it as home. I was born here. My father and his father were also born here."

"Then why isn't it called Kensington Castle."

Jonathan laughed again making Brian feel a bit silly and naive. "There is a Kensington Palace that belongs to the King. My family bought this castle and the land nearly eighty years ago. The name existed at least a thousand years before that, so it just didn't seem right to change it. Even though my family owns the place, it is part of the country's history and heritage. It just wouldn't be right to change it."

"Since you put it that way, I can understand," Brian responded with lightness to his words.

"Brian," Jonathan began, "I want us to be friends because of who we are, not because of my family, or my home, or anything else. I don't want you or our friendship to be affected by any of this."

"Ah, don't get so serious on me. While you cannot help the impression part, my parents raised me to see the person behind any facade. You are a good man, a great pilot and a splendid friend. Don't worry about me."

Jonathan leaned forward opening the sliding window to the driver's compartment. "Johnson, step on it. You've got two thirsty pilots who need a few pints of ale."

"Yes, sir," Johnson answered speeding up slightly without adding risk.

The automobile passed from the forest into an open, hilly, green field carved out among the trees. As they moved up the closest rise, the top of Carlingon Castle came into view. The building of modest size, by castle standards, situated upon the adjacent hill and the highest point defied Brian's imagination. The circular granite gray tower complete with notched parapet dominated the structure. A more conventional stone building of later vintage had been added onto the original tower. Plumes of white smoke were carried away sharply from several chimneys. All the windows within view had noticeable depth to them and were decorated with delicate, white lace. A massive, deep brown, weathered, oak door with black, wrought iron straps dominated the front of the castle. Johnson took the clockwise route around the circular drive opening onto a large rectangular park area that extended the length of the castle.

The door opened as if by magic as they approached. "Good afternoon to you, Mortimer. A pleasure to see you as always."

"Good afternoon, sir," answered Mister Mortimer, the house butler and chief servant.

"Mortimer, this is my best friend, the American I told you about, Pilot Officer Brian Drummond."

"An honor to meet you, sir. Mister Kensington speaks quite highly of you. It is my pleasure to welcome you to Carlingon Castle."

"Thank you. It is nice to meet you."

Looking to Jonathan, Mortimer continued, "Your father and mother are in the study. Miss Rosemary has informed us she will not arrive until late tonight or tomorrow."

"Very good," Jonathan responded then turned to Brian. "Rosemary is my younger sister. She is struggling through Oxford. Let's go meet my parents. I know they are eager to meet you as well."

Brian followed Jonathan down a long hallway that appeared to run the length of the house from the foyer to a large room, presumably the living room, at the far end. The floors, walls, ceiling and large, solid beam ceiling supports were all dark, weathered oak. Modern incandescent lighting made the hallway glow a golden color. An open door halfway down the hall on the left provided access to the study. Virtually every part of the study walls, from floor to ceiling, except for the window in the far wall, were covered by books. A sliding ladder was installed on each wall.

Mister George Kensington rose from behind a commensurably large, rich red, mahogany desk with green leather top cover. Missus Theona Kensington sat behind a table top loom in the right corner of the room.

"Great to see you again, son," Mister Kensington said. Before they could shake hands, Jonathan's father continued, "This must be the famous Brian Drummond we have heard so much about." He shook Brian's hand with understandable firmness and warmth.

The get-acquainted words passed among them. Brian took an instant liking to the warm, amiable and caring parents. In many ways, they reminded him of his own parents. Missus Kensington worried about Brian's mother's feelings, concerns and worries, commiserating with a woman she had not met, based solely on the service and sense of duty of their sons.

The house, the castle, even though it was quite old, was filled with laughter and enjoyment of life despite its hardships. The evening meal came and went among the questions, answers and discussion common to so many household tables in Europe . . . the events of war. The appreciation of Churchill's courage over the previous decade was strong, genuine and profound, and strengthened Brian's connection with the Kensington family.

Comments about their daughter, Jonathan's sister, left Brian with an impression of a headstrong, intelligent, accomplished and confident young woman only one year older than himself. He respected her commitment to Oxford University even though it was not fashionable for women to attend a university. According to the description, she intended to become a doctor of medicine. They were without question proud of her although sometimes baffled by her stubborn pursuit of personal objectives. Brian looked forward to meeting Jonathan's sister.

―

Sunday, 24.December.1939
Carlingon Castle
Newcastle-upon-Tyne, Tyne & Wear, England

"Merry Christmas, Brian," Missus Kensington said as he entered the living room. The large fire provided a deep warmth to the room. No one else seemed to be up. Brian saw the moderate size, near perfect, conical pine tree decorated with hand crafted ornaments and a foundation of brightly colored, ribbon bedecked presents, and crowned by a golden star. The symbol of Christmas produced a smile and a shot of home sickness. Brian missed his parents, and for the first time in many months, he remembered Becky and the enjoyment they all had at this time of year. The celebration of the birth of Jesus Christ always brought a sense of renewal, of rebirth, as he had been so carefully taught by his parents.

"Merry Christmas, ma'am."

"I trust you had a good night's sleep."

"Yes. Quite good, actually. I don't usually sleep this late, but we had quite a few long patrols this week. I suppose the fatigue was catching up with me."

"Quite understandable. You are certainly not the last to rise. It would appear Mister Kensington and Jonathan are later than you."

Brian assumed Jonathan's sister was not yet home. Missus Kensington talked about the day's events, as well as asking Brian about his desires and religious beliefs that seemed to be coincident with theirs. The family would have a light meal when the other two men of the house rose from their comfortable slumber.

Theona Kensington presented an interesting mixture of traits. She was outgoing and yet quite reserved, sensitive while equally strong, intelligent but not overbearing. She was certainly proud of her children although she did admit being very worried about Jonathan's interest in flying which was born at an early age. The prospect of war and her only son facing what her brothers endured in the killing fields of the Somme brought out a strain of resentment in her. As Brian listened to Theona Kensington's thoughts about the Great War, her brothers' experience and the risk to her son, he could not help thinking of his own mother's worries about him. Letters from his mother told him the two mothers on different continents with only sons serving as fighter pilots shared a great deal although they had never met. Brian vowed to write an understanding letter to his mother.

The day progressed peacefully, slowly and graciously. An early afternoon horseback ride left Missus Kensington behind to wait for her daughter's arrival. It also presented Brian with the differences between the hornless English saddle and the American saddle he was accustomed to. A quick lesson in English riding technique from Jonathan and some initially timid riding helped Brian make a quick transition.

The beauty of the countryside despite the chilly temperature enthralled Brian. The contrasts with his native Kansas were dramatic. The sky cleared overnight leaving small, puffy cumulus clouds drifting slowly eastward on the light breeze. The expanse of green in all directions and the variety of colors, shades and tones still staggered Pilot Officer Brian Drummond. The three men stopped several times to inform Brian about the significance of one view or another, and help Mister Kensington understand the importance, the thrill, the joy of flying, and more specifically flying the renown Supermarine Spitfire. The thought came to Brian several times during their discussions that George Kensington was gathering information to influence his wife or at least help her deal with the duty of her son. He wondered if his own father or maybe Malcolm Bainbridge were doing the same thing for his mother.

The three riders returned to Carlingon Castle as the afternoon light diminished with the sun dropping below the mountains to the west. It was actually half past three with the short winter days of the Northern latitudes.

A stable hand tended to the horses which Brian benefited from for the first time in his life. The young man, maybe slightly younger than Brian, was the son of the butler.

Brian followed George and Jonathan Kensington into the kitchen where Theona Kensington and three large, steaming mugs of hot chocolate greeted their return. Missus Kensington told the gathered family members, guest and kitchen staff that Rosemary had returned over an hour ago and was upstairs resting. As told by her mother, Miss Kensington had attended an end of term party that continued on the train north from London. Her father did not approve, but also shook his head in parental resignation over the antics of his headstrong daughter. The intriguing image of Jonathan's sister fascinated Brian. He wondered what she was really like.

Christmas Eve dinner would be served in less than an hour. Everyone retired to freshen up before the evening meal. Families came together for Christmas Eve in anticipation of one of the most celebrated days in Christendom. Brian did not need to be told it was not only unusual, but a bit of an honor to be included in what would otherwise be a family occasion.

Jonathan and Brian, the first to arrive in the large, high, oak beamed ceiling, living room, wore their RAF blue uniforms proudly. A well-prepared fire in the enormous fireplace added a pleasant glow to the room adorned with awards, memorabilia, photographs and several ancestral paintings. Mister and Missus Kensington entered the room together. He wore a dark suit and tie while she wore a very complementary emerald green dress which had a deep richness to it. Mister Kensington directed the sherry to be served since there was no indication when or even whether their daughter would join them. The smooth, woody drink brought additional warmth to Brian that he had grown to enjoy since arriving in England six months ago.

Jonathan decided to change from small talk about the weather and economic prognosis. "Brian is a smidgen on the famous side. The uncle of

Brian's RAF benefactor is none other than Winston Churchill himself."

"Is that so?" asked George Kensington rhetorically. Brian nodded his slightly embarrassed confirmation.

"I'll be damned." A sharp look from Theona Kensington corrected him. "Excuse my poor language. I'll be . . . ."

"Not only that father, he has met Mister Churchill and had a long, friendly, private chat with the man."

"This cannot be so?"

"It is. He went to see him at Chartwell, no less."

"How did you manage that, Brian?" asked Theona Kensington.

"The man who taught me how to fly was an American volunteer with Four Three Squadron during the Great War. His best friend was, and still is I suppose, John Spencer, who is related to Mister Churchill. Group Captain Spencer is currently the staff secretary to Air Chief Marshal Dowding, the commanding officer of Fighter Command."

"My, my, I am impressed."

At that moment, one of the most strikingly beautiful women Brian had ever seen walked into the room wearing a low cut, off the shoulder, bright red, floor length dress that accentuated her modest but well-shaped bosom. Her moderate length, wavy, blond hair and sky blue eyes immediately reminded him of Virginia North, Anne's best friend and Jeremy Morrison's lover. Rosemary was not as well-endowed as Virginia, but her facial features were more elegant, porcelain and attractive. Without acknowledging anyone else in the room, she walked directly to Brian who was now standing and looking directly into the woman's eyes.

"My, my, indeed, Father," she said stopping within arm's reach of Brian making him feel very uncomfortable. "I too am impressed." While continuing to look deep into Brian's eyes, she said, "Well, brother of mine, who is your friend?"

"Brian Drummond, this is my younger sister, Rosemary. Rose, meet Brian."

She extended her right hand for a shake rather than the continental kiss. Brian grasped her hand firmly but gently. Her hand was warm, dry and soft. She held him even though he had released her hand making Brian even more uncomfortable. Her closeness by itself did not bother Brian, but her closeness with her older brother and parents watching them brought a slight flush to his cheeks.

"How absolutely delightful. I seem to have embarrassed our guest."

"Rosemary Alice," chided her mother, "get control of yourself."

"Oh Mother, I mean no harm or discourtesy. Mister Drummond is simply an excellent specimen of a man and a defender of our honor."

"Yes, however . . ."

Rosemary held her hands up as if she were being held up by a robber and backed away from Brian. She accepted her own glass of sherry from Mortimer. Brian looked away from Rosemary for the first time since she entered the room. He received apologetic expressions from her parents and a wink from Jonathan.

"For your benefit, Rosemary, prior to your entrance, Brian was telling us about his meeting with Winston Churchill."

"The war monger," challenged Rosemary obviously baiting her parents.

"Rosemary, stop."

"My apologies, Mister Drummond. I pray I have not offended you."

"No offense taken," Brian responded looking into her seductive eyes.

"What does Mister Churchill think about the situation in Europe?" asked George Kensington purposefully changing the subject.

"I met him at the beginning of August, before the war began. Even at that time, he thought general war with the Germans was inevitable. I would say his premonition, or wisdom, has been vindicated."

"Does he think Great Britain will be drawn into a land war?"

"From what I recall, he feels that is equally inevitable."

"This is Christmas," protested Rosemary.

"And, it may be the last Christmas of semi-peace we may enjoy for some time," added George Kensington.

"The Royal Navy certainly would not call it, peace," interjected Jonathan entering the conversation for the first time.

"At least they got that German battleship," mused George.

"The *Graf Spee*," added Brian.

"That's it."

"They certainly did get it," answered his son. "And, the Finns have done rather well, fending off the Communists."

"We may not be in the best situation, but if the Finns can do it against the Red Army, well . . . anyway . . . we shall prevail, the Allies shall be victorious."

"Most of our family and friends tend to agree with Churchill, now that the Germans have invaded Poland," observed Theona Kensington.

"He will be the prime minister, soon," mused George Kensington.

"Heaven forbid," said Rosemary.

"Enough politics," Theona Kensington said. "This is the eve of our Lord's birth, and we shall have an enjoyable evening together. Mortimer, if you would be so kind, we will take our supper, now."

"Yes, madam," the butler responded.

The dining room equaled the remainder of the house rich in stone and woods. Mister and Missus Kensington sat at either end of the table, and Jonathan and Rosemary sat together opposite Brian. The five course meal prepared with care and served with precision complemented the conversation which for the most part remained light and jovial. The only disturbance for Brian occurred under the table when Rosemary repeatedly touched him with her foot and a wink or a smile. Brian felt a strong attraction to her physically although he remained concerned for Jonathan's feelings. His friendship with his squadron mate was more important than a night's pleasure with a beautiful woman. Unfortunately, he was reminded numerous times it might not be up to him. A feeling of being cornered began to creep into his consciousness which drove him to ignore Rosemary. The more distance he tried to create, the more determined she seemed to become.

As they retired to the living room to converse about tomorrow's events, Brian recognized the effects of an abundance of wine and passed on the brandy. He made a concerted effort to avoid Rosemary's eyes that he sensed numerous times were on him. Her alluring appearance made his resistance difficult, but he also knew if he got any closer to her, he would be drawn into her web. The assault on his senses was too much. He was inwardly very grateful when Jonathan began the cascade of movements toward the magnet and safety of sleep.

With the door closed, curtain drawn, and lights on in his small, but well-appointed bedroom, Brian had only removed his tie, jacket and shoes when a soft knock came to his door. Instinctively, he knew who it was. She stood before him without one item of clothing or even hair amiss with the most devious smile and expression he had ever seen.

"I know you will think this bold, but I understand Americans like the direct approach," she said. "I have been immensely attracted to you from the moment I saw you. I want you tonight. I want to know you far better than I do now. Come to me tonight. In an hour, everyone will be sound asleep, and we can make all the noise we want."

The thought of making love to a beautiful woman who desired him forced him to consider the possibility. Jonathan, the Kensingtons and Carlingon Castle bolstered his restraint. "I can't."

As if she did not hear his response, Rosemary whispered, "Are you going to make me stand in the hall for everyone to see?"

Brian's thoughts raced from one extreme to the other, and from one possibility to another. He stood back to let her into his bedroom. As soon as she pushed the door closed, Brian knew he had made a mistake.

"You do not have to worry about a thing, Brian."

"I already have a girlfriend."

Rosemary snickered. "How quaint! Brian, I am not asking you for a long-term commitment. I am simply intrigued by your size. I want to experience the rest of you. If something more develops from our simple carnal pleasure, then all the better. What I need is a good fucking."

Brian fought to keep from reacting to the strongest, most forward language he had ever heard a woman speak. He did not want to seem young and naive, as he knew he was. "Your brother, my best friend, is in the room next door. Your parents, my hosts, are just down the hall. I am very grateful for the family's hospitality, and I don't want to offend anyone. Anyway, it is late and we have an important day tomorrow."

"I hope you recognize your prudishness adds to my desire."

The confines of a tight space with a predator salivating before him drew tighter. Brian felt more cornered than he had ever felt in his life. Problems in the air paled in comparison to the incredibly awkward situation he currently found himself in. He wanted breathing space and yet he felt powerless before this attractive, strong and confident woman. His relief came in the form of Rosemary's timely resignation.

"As you wish, then. I shall leave you alone for tonight, but remember, it is not as warm sleeping in an empty bed." Rosemary Kensington did not wait for a response. She left the room shutting the door quietly behind her leaving Brian standing in the middle of the room absolutely dumbstruck.

Monday, 25.December.1939
Carlingon Castle
Newcastle-upon-Tyne, Tyne & Wear, England

THE day began with a rushed light breakfast in order to arrive at the small community church in time for Christmas morning service. Although Brian, at times, had difficulty understanding the sermon, the service replicated Christmas church celebrations in Wichita, Kansas. The day's weather rapidly set the tone for the day as the overcast burned away leaving a bright crystal blue sky and unusually warmer, although still cool, temperatures.

Rosemary Kensington kept up her flirtation with Brian although she did try to be more discreet. A large English Christmas meal was served in the early afternoon offering new gastronomic enjoyment for Brian. The entire Kensington family and Pilot Officer Brian Drummond used the remaining daylight for a ride through a portion of the countryside Brian had not seen yesterday. Several times during the ride, Rosemary tried to lure Brian away from the others. As with the night previous, Brian was sorely tempted, but resisted for the same reasons.

Sunset ended their ride. A quick freshen up and change of clothes preceded a light meal of cheese and fruit. While the family relaxed in the living room with a robust fire, the traditional exchange of gifts embarrassed Brian again since he had been unable to obtain any gifts which Jonathan had assured him would not be expected. Nonetheless, Brian felt quite odd with gifts passing between the family members. The Kensington family gave both Jonathan and Brian an elegant, wool, RAF blue sweater to keep them warm in the cold of high altitude flight.

As the large clock above the fireplace mantel struck 21:00, George Kensington rose from his chair to turn on the radio in the corner of the living room. Without looking back, Mister Kensington simply said, "It is time for the King's Christmas message."

The room fell silent. The assemblage listened carefully and intently to the clear, confident, but solemn voice of King George VI. The words were not particularly happy given the world events around them, but it was a message of hope. The King closed his annual message with, "I said to the man at the gate of the Year, 'Give me a light that I may tread safely into the unknown.' And he replied, 'Go out into the darkness and put your hand into the hand of God. That shall be to you better than light, and safer than an unknown way.' May that Almighty Hand guide and uphold us all."

The room remained silent except for the crackling of the fire. Everyone seemed to be absorbing the King's words and what they meant to them. Brian had heard President Roosevelt speak on the radio several times about the need for community commitment and confidence to assist recovery from the Great Depression. Although Brian had not experienced the rigors, finality and hardship of combat, yet, he had a fairly good image of what to expect. The King's message, in the light of his image of what was near and lay before them, firmly placed a lump in his throat. He knew it was a simple message, but most impressive and poignant.

Eventually, each of them returned to the room. The discussion shifted rapidly to the perceptions of danger associated with flying and the possibilities of combat in the skies over Great Britain. Jonathan and Brian told the others about the thrill of speed and power manifested by the Spitfire and of the challenges of the German reconnaissance flights. The fact that German aircraft were routinely over-flying the United Kingdom bothered the others, stimulated their anger and pride, and brought home the danger faced by their son and his friend.

"We shall have a slow Boxing Day," Theona Kensington said as a subtle hint to conclude the day.

"Yes, very good, dear. We shall have a slow Boxing Day."

"What's Boxing Day?" asked Brian.

The soft chuckle from each of the Kensington's conveyed a friendly message. Missus Kensington answered. "It's a traditional day in England dating back more years than I can remember when we box up items to offer to the poor and less fortunate than us."

Brian Drummond liked the idea. The hour turned late and the group disbanded. Brian quickly prepared for bed. He was tired. The initial cool, softness of the sheets would feel so good, soon to be followed by the enveloping warmth of the thick *duvet*. With the lights of his room extinguished, Brian drew back the heavy curtains to appreciate the blanket of stars and the rising half-moon. Only a few clouds occupied the sky. He stood before the window.

Brian's thoughts fluctuated between the soon to be experienced combat and the words of King George VI. He gained some comfort that a man as important as the King recognized the difficulties the nation would soon face. At least, they would not go into combat alone. He felt better for the King's words. The action of slipping into bed brought an immediate relaxation with sleep soon to follow.

The soft squeak of the door hinge brought Brian back to awareness. As he looked up, he saw Rosemary gently closing the door behind her. His heart jumped as she turned around. The dark triangle and two circles on her chest were clearly visible through her thin white gown. The soft diffuse moonlight illuminated her body and smile as she approached the bed. The young pilot did not know whether to accept what he thought was coming, or to continue his resistance in the name of family relations. His resistance began to crumble as she moved slowly toward him. Brian quickly got out of bed thinking if was his only hope.

"I have come for you," she whispered as she stopped within arm's reach.

"Excuse me."

"No need to be coy with me, Brian. You pilots are all alike. You want this as much as I do."

"Rosemary, please. As I said before, I have a girlfriend," Brian overstated his relationship with Anne. "Your brother is my best friend, and we are in your parent's house."

"Not to worry," she said as she reached for the part of his anatomy that most interested her.

Brian felt her hand grasp him causing a reflex jump back away from her. "Whoa."

Rosemary advanced toward him. "Relax, Brian. Your body is speaking for you. Just let it happen. No one will know. You want this, too." Rosemary Kensington dropped the thin material of her gown. Brian could not avoid absorbing the soft curves of her hips. Her trim, shapely body offered more than he could bear. The dark, conical nipples on her modest breasts drew his hands to them. As he felt her firmness, the smile grew as if she knew she had won the small battle. Their lips joined completing her victory. The passion between them grew rapidly. She stopped to pull back the sheets, and then pulled down his pajama bottoms.

Looking down, Rosemary caressed him. "Oh my. You are indeed every bit as much as I imagined. We are going to have some fun." She grinned even larger as she guided him back to bed. Rosemary giggled audibly as she jumped up on the bed, lay back and raised her knees wide. "Come to me, Brian. Come make me a happy woman."

Any semblance of resistance vanished in a flash. Brian hopped to the task before him with unbridled enthusiasm. For this moment in time, there was no other world beyond her embrace. He used all the lessons learned with Anne Booth and experienced a few new facets of human sexuality. Rosemary Kensington brought humor, fun and light-hearted enjoyment to sex. The two lovers took the fullest advantage of one another with the insatiability of youth until exhaustion enveloped them both.

Chapter 18

War is like love, it always finds a way.
-- Bertolt Bercht

Friday, 5.January.1940
RAF Drem
Drem, Lothian, Scotland

"I hope you two blokes had a good time during your holiday," announced 'Mongo' Strickland as 'Harness' Kensington and 'Hunter' Drummond entered the Dispersal building an hour later than the other squadron pilots. "The skipper wants to see you two." They both knew why.

Brian knocked on the door to Squadron Leader 'Spike' Darling's small office.

"Enter," he said with a noticeable irritation. The two young pilots entered and stood at attention. Darling continued to work on his paperwork for several minutes making the two tardy officers wait. When he was ready, he put his pen down and stared at the two men. "You two have enjoyed two weeks of holiday in the middle of a damn war, and you cannot seem to make it back here on time." The two men knew there was no excuse. "What have you got to say for yourselves?"

"We have no excuse, sir," answered Jonathan.

"And what about you."

"They had a problem with the train, sir."

"They had a problem with the train, did they now. I suppose old Adolf himself bombed the tracks."

"No, sir," Brian responded wishing he had not tried to answer his leader.

Darling stared at his two newest pilots making them feel even smaller. He stood up from behind his simple, small, wooden desk to walk slowly behind the two pilots. "You are required to be at your station at or before your duty time. There are no, I repeat NO, excuses to relieve you of that responsibility. You are essential elements of a team vital to the defense of this country, and I will not permit anyone to degrade the capability or performance of this team." He paused presumably to let the impact of the words sink in. They did. Brian did not regret his last union with Rosemary in the dim light of the pre-dawn, but he did regret leaving for the train station late. "We are on a 30-minute alert. I suggest you two should gather up your flying kit and join your mates. We will not have this conversation again. Is that perfectly clear?"

"Yes, sir," the two men said in unison. Squadron Leader Darling waved his hand dismissing his two pilots.

As they joined the others, silent snickers, raised eyebrows and other gestures of friendly gibes greeted them. The good natured ribbing continued as Brian and Jonathan sat in available chairs with their flying clothing and equipment next to them. Corporal Jennifer Warren added their names in the appropriate positions on the status chalk board. The telephone rang instantly terminating the jovial conversations and drew all eyes to it. The operations clerk lifted the handset and listened. The message was quite short. She placed the handset in the cradle as 'Spike' Darling joined his pilots. The corporal announced, "Squadron to readiness."

"There we go, lads," Darling said. "Looks like we may have a customer."

Each pilot donned their winter flying clothing, fleece lined, leather trousers and jacket with similar boots. May West floatation vests were either donned or in hand along with the leather flying helmet, headset, oxygen mask and gloves. Several pilots decided to walk casually toward their fighters. Brian and Jonathan joined that portion of the squadron. It was a chilly, overcast day fortunately with only a light wind. The pilots began to disperse without words as each prepared for a combat patrol not quite imminent. Brian saw the chunky frame of Leading Aircraftman Gordon completing his preparation of the 'Freddie' bird's cockpit.

"Good morning, lads," Brian said to his crew as he approached the elegant Spitfire poised for takeoff.

"Good morning, Mister Drummond," all three crewmen said virtually in unison.

"Your mount is ready, sir," Gordon added knowing they had only been brought to readiness.

"Thanks, Bernie." When Leading Aircraftman Gordon jumped off the left wing, Brian ascended to the cockpit connecting his communications cables and placing his helmet and oxygen mask over his pipper gun sight. The gloves went to the top right of the instrument panel. The pilot checked his parachute harness and seat straps, all in perfect order as they always were, thanks to Leading Aircraftman Bernard Gordon.

"How was your holiday, Mister Drummond?" asked Aircraftman Colin Jenkins, Brian's armorer. The ground crew instinctively knew not to bother the pilots once they had gone to standby status. Each pilot had his own way of dealing with the uncertainty of aerial combat and the proximity of death or injury. Readiness status allowed the pilots to be a little more relaxed. Brian Drummond had always been fairly relaxed.

Jenkins question brought instant, vivid recollections of Rosemary, her delightful and athletic body, and the unmitigated enjoyment of their encounters.

It also brought clear memories of Anne Booth, her diminutive, full body and worldly experience in matters of the flesh. Even a thought of Rebecca Seward and their exploratory innocence floated among his bountiful memories. "I had an absolutely brilliant time. I spent the Christmas holiday with Mister Kensington's family outside Newcastle."

"So, you got to experience a good English Christmas, did you now?" asked Leading Aircraftman Gordon.

"Yes, indeed, and it was my best ever," Brian responded with an enormous grin on his face knowing his ground crew had no way to understand his true meaning.

"Squadron to standby," announced Corporal Warren as the remainder of the pilots jogged to their aircraft.

Brian climbed into his machine. They were now minutes from launch on a combat patrol. Something was out there. They did not know what, nor did they have any way to find out other than interception. Bernie Gordon helped him with his parachute harness and seat straps. Before Brian finished his cockpit procedures, the scramble bell was ringing. He was not the first to get his big Merlin kicking. As a well-oiled, powerful machine, the entire squadron was airborne before Brian began to consider possibilities.

"Rooker, Sorbo is airborne," radioed 'Spike' Darling to Turnhouse Sector Control with his full, twelve aircraft squadron climbing to the East at full power, +3 inches of boost.

"Roger, Sorbo. Climb to angels one eight, heading zero six seven. Ten plus bogeys inbound at angels one five."

At that moment, each pilot silently knew combat was now closer. They armed their guns. The cockpit was readied for combat. For Brian, this was the closest he had come yet to firing upon an enemy aircraft in anger. There was not much satisfaction in the knowledge the German raid would have no fighter escort due to the range from Germany. The altitude of the incoming raid signified a probable bombing mission. The reconnaissance missions were much higher, usually. The He111 had already demonstrated the lethality of its guns on careless fighters. He did not want to be another victim to the enemy, especially on his first engagement.

"Sorbo, this is Rooker. Change heading to zero seven one."

"Sorbo, roger."

On this winter day, the layers of broken clouds seemed endless. Brian kept his eyes glued on his section leader, 'Jackstay' Beamish, and did his best to take a quick glance beyond his leader each time they passed between layers. He wondered how they would ever find the enemy aircraft among all the clouds.

The flight of twelve aircraft leveled off at 18,000 feet in a cloud layer.

"Rooker, Sorbo. We are at angels one eight in the klag, heading zero seven one."

The pause in response certainly indicated the effort of the sector controller to determine the best correction to make. The squadron of fighters would undoubtedly not be able to find their target in clouds. "Sorbo, this is Rooker. Turn to one six zero and descend to angels one seven or until below the cloud layer."

"Roger, Rooker." The entire flight followed their leader. They were not out of the cloud at 17,000 feet as they continued to descend slowly.

The frustration, growing tension of flying in close formation, in and out of clouds, and knowing an unseen enemy lurked among the clouds made beads of sweat descend down his back in the cold cockpit. The vectors and altitude changes continued for nearly an hour. Frustration mutated toward anger. Brian's muscles and joints ached from the tension. Finally, relief came.

"Sorbo, this is Rooker. If no joy, return to base."

"Rooker, no joy."

"Sorbo, descend at pilot's discretion to angels three, heading two seven zero."

"Roger, Rooker."

The flight broke out of the clouds at 5,000 feet over the North Sea. The coastline appeared several minutes later. As the details of the coastline became more evident, Brian recognized they were southeast of Drem. Darling adjusted their heading accordingly. Brian noticed four tall, trellised towers with large rectangular open structures on top. A small wooden house sat in the middle of the towers. That must be the radio direction finders some of the guys called, RDF, or sometimes Chain Home, Brian said to himself. As they flew low, below the tops of the towers, Brian wondered how they worked. Somehow they used radio waves to find airplanes in flight. The operation of the mysterious system stimulated Brian's inquisitive nature.

Landing at RAF Drem was uneventful. The gun port patches were still in place on his wings. His ground crew waited patiently for Brian to turn the Spitfire and shutdown the engine. The duration of the flight and the presence of the gun patches told the crew everything they needed to know about the flight.

"Snags, sir," asked Leading Aircraftman Gordon.

"She flew like the lady she is, Bernie," answered Brian as he jumped down from the wing feeling the shock on his cold legs and stretched.

The intelligence debriefing took only a few minutes. They searched, they found nothing and they came home. Darling called Sector Control to find out more about what they believed was out there and to tell the controller what happened in the air. Only half the conversation did not help much.

"Well, lads, the raid did not make it either. We are at thirty minute alert."

The pilots gradually began to come down from the anxiety of the flight. Brian melted into a chair without removing one item of his flying kit. He had many questions on his mind but waited for the conversation of pilots to take over.

"'Jackstay,'" Brian said and waited until he had the attention of Flight Lieutenant Roger Beamish, "were those towers as we crossed the coast the Chain Home you've talked about?"

"They were indeed, laddie. A good, close-up view, I'd say."

"How do they work?"

"They send out a radio wave that bounces off an aircraft. They pick up the return signal and somehow figure the rest out."

"You've said that before, but how can they tell height, direction and other stuff?"

"You are past me, 'Hunter.' I would suggest you talk to the Chain Home blokes. Maybe they will tell you how they work?"

"They are not going to tell him how it works," interjected 'Mongo' Strickland. "It is supposed to be secret."

"Our lives just might depend on that system and for all we know our ability to defend these islands may rest directly upon the Chain Home system," 'Spike' Darling added.

"Skipper, can you get me a pass to visit that place?" asked Brian.

"First, that place, as you say, is called, Drone Hill, and I'll see what I can do."

"Thank you, sir."

―

Wednesday, 17.January.1940
RAF Drem
Drem, Lothian, Scotland

"SCRAMBLE, Blue Section," came the command as the telephone handset descended to its rest. The others either remained in the Dispersal building or stood just outside to watch 'Spike' Darling lead his two wingmen, Pilot Officer Roland 'Boxer' Stockard and Flying Officer George 'Angle' Ashcroft, running toward their poised Spitfires.

"Two minutes, fifteen seconds. Not too bad, I'd say," observed Flight Lieutenant Robert 'Sparky' Morrow, the Red Section Leader, Operations

Officer, and number two leader of the squadron, as Stockard, the last of the three fighters, lifted off. All three Spitfires banked hard as their undercarriages retracted into the wings climbing hard at full power.

"Damn, but those engines sound good," said Brian Drummond more to himself than to anyone else.

"Don't hold back, now, 'Hunter,'" joked 'Mongo' Strickland. The assemblage laughed as the three Spitfires disappeared into the overcast, although the melodious tones of the Merlin engines were still clear. Everyone returned to the relative warmth of the Dispersal building and the activity they did most – waiting.

Brian looked to Roger Beamish, got his attention and leaned forward. "How do the controllers give us directions on where to be?"

"My, aren't you the curious one."

"I just figure if I know how the system works, I can work better within it."

"All you need to do is find a bloody German and shoot his arse down," interjected 'Sparky' Morrow. Those that were listening laughed or nodded their agreement.

"Well?" Brian returned to his question.

"If you are going down to Drone Hill, why don't you go up to Turnhouse Sector Control? I am certain they can answer your questions."

"Where is Turnhouse?"

"At RAF Edinburgh, west of the city."

Brian considered when he would be able to visit RAF Edinburgh as Beamish suggested. He also wondered if he would need permission for that visit as well. "OK, but do you know basically how it works?"

After some jibes and kidding of Brian Drummond by several of the other pilots, Beamish responded, "As I understand the process, Sector Control receives target indications from Chain Home or the Observer Corps. They must plot the information, determine what response is required and scramble the appropriate number of fighters to deal with the threat. They keep track of all the targets and all the available fighters. I am certain they must have methods to regulate the refueling and rearming procedures to ensure invading bombers don't feel neglected by our fighters."

"Have you ever seen this control system work?"

"Yes, laddie. Most of us have. It works. You will see for yourself. We just have not had much business up here in the North, although what action there is seems to be up here."

"Do you think the Germans will invade in the North?" Brian asked remembering the discussions with his colleagues at RAF Hawarden.

"Aren't you full of questions? Maybe you should go back to school," Beamish said more for the others than for Brian. Most of the other pilots did not miss the opportunity to jump Brian for his inquisitiveness although he knew some of them wondered about the details of the system as well as he did. Brian was not the youngest pilot in the squadron. Flying Sergeant Miles 'Fog' Johnson, the quiet, average built, Yorkshire, aviation enthusiast, was nearly two years younger than Brian. The American was simply a more highlighted target of jokes and kidding.

The distinctive melody of the Merlins faintly returned to the Scottish hills along the Firth of Forth. "That was a rather quick patrol," observed 'Organ' Foxworth.

"The patches are gone," someone outside yelled.

"No wonder it was quick. Looks like the Skipper may have had good hunting. Wonder if he bagged anything," added Foxworth.

Brian looked out the window with Jonathan Kensington alongside. The squadron leader's aircraft taxied to the spot closest to the Dispersal building. The distinctive red cloth patches were indeed missing with the tattered remnants waving with air passing over the wing. It was the first time either of the young pilots had been close to a fighter that had been engaged in actual aerial combat. Brian felt a measure of excitement as well as realization. He knew he was closer to his dream and yet the closer he became the more stark the dream became. There seemed to be an ominous finality to those missing gun port patches.

They watched the crews scurry about like ants tending to their queen refueling and then rearming the three fighters. The impression left while observing the frenetic activity was one of a task left undone although none of the other sections had been placed on readiness. The three Blue Section pilots returned to Dispersal.

"Good hunting, skipper?" asked 'Jackstay' Beamish.

"Yes, indeed. We bagged a Heinkel One One One."

"We, hell," interjected 'Angle' Ashcroft, the Blue Section right wing, "the skipper did the bloody Krauts in, on his first pass. Into the sodding sea, no chutes."

"Congratulations, 'Spike,'" said Beamish. Congratulations from the other pilots followed. The He111 was the first victory for Squadron Leader Darling and the first for the squadron since the war began. General euphoria and elation occupied most of the squadron that winter Wednesday morning. Brian could not erase the words 'no chutes' from his consciousness although outwardly he joined the celebration. Four human beings died this morning.

The rough edge of his dream inched closer.

—

Wednesday, 17.January.1940
The Admiralty
Whitehall, London, England

WAITING without words or expression in the First Lord's anteroom were Colonel Stewart 'C' Menzies, Vice Admiral Sir Geoffrey 'Jumper' Pike, William 'Intrepid' Stephenson, and Alastair Denniston, Head of the Government Code and Cypher School. They all knew why they were there waiting to talk to the First Lord of the Admiralty. There was no need for conversation to occupy the silence.

"The First Lord will see you now," his secretary announced.

The four intelligence professionals moved swiftly and quietly into Churchill's office. Cordial greetings were dispensed with promptly and properly.

"What do we have?" asked the First Lord to begin the private and most secret meeting.

'C' looked to Denniston. "First, I should remind everyone, the conversation from here on is, Most Secret-ULTRA." Everyone nodded. "Alastair, if you would be so kind."

"Through a series of contacts in Scandinavia, and some careful and protracted discussions, we have obtained a copy of a multi-level set of codes for both the Russian and German military from a critical and reliable source within the Swedish military. He has verified the Swedish effort to validate most of the codes. They have been successful with most of the Russian codes. There are several of the German codes they could not validate, but they believe them to be active."

'C' nodded for Denniston to continue.

"First Lord, we have performed a quick assessment of the codes our benefactor indicated they have not validated. Checking the codes against our knowledge of Enigma has given us a positive estimate they are most likely Enigma sequences. It will take us another few days to several weeks to fully evaluate the codes. If our estimate is validated, these sequences will vastly improve our productivity for ULTRA."

There was no smile on Churchill's round face. "We must thank the Lord at least something is going our way."

The other men knew the First Lord of the Admiralty felt every loss at sea. While the Royal Navy bore the brunt of engagements with the Germans, Churchill did not take even minor losses or setbacks very well. While Admiral Pike knew personally the burden the First Lord carried, the others sensed the additional weight although unsaid.

Denniston continued. "We are taking the appropriate German codes under ULTRA protection. The remainder of the codes vary from low level field codes to fairly sophisticated high level codes not associated with Enigma. The GCCS will retain all the codes to avoid any linkages but retain them under different file names. The value of these codes from Sweden is incalculable."

"Did the Swedish government sanction this gift?" asked Churchill.

"We cannot be certain," responded Menzies. "Indications received from our contact plus a few indirect bits from our community would lead us to believe there was high level support for the exchange."

"We should discreetly thank the Swedes."

"With respect, Winston," said 'C,' "I don't think that would be a good idea. While we are immensely grateful for these codes, we cannot risk any possible inference the codes may be more valuable than the Swedes already know."

"I might add, First Lord," added Denniston, "our contact was well aware of the significance of the keys by themselves. I do not believe they have even the slightest inkling we have an Enigma device."

"Do they expect any reciprocity?"

"No, sir. These codes have been provided with only one condition . . . they should be used to defeat Hitler's expansionist ambitions. The Swedes feel the heat. They suspect it may be only a matter of time before their country is attacked by the Germans, or Russians, or both. While they have given us use of the codes without technical conditions, they intend to utilize the codes to help the Finns against the Russians, and the Norwegians and Danes against the Germans. In fact, the Swedish codebreakers determined the Russian invasion plans of Finland including the invasion date of the 30th of November. The information was provided to Field Marshal Baron Carl Mannerheim leading the Finns who, as you know, held off the Russians against incredible odds. The Swedes supplied a steady flow of timely, decoded, Russian messages."

"The Swedes are meticulous with materials like these codes. This arrangement should be a minimal risk to our exploitation of the codes," said Admiral Pike anticipating the probable next question.

"First the Norwegians. Now the Swedes," said Winston. Pike and Menzies exchanged brief looks of concern, and then a quick glance at Stephenson and Denniston. Churchill did not miss the implications. "Have I misspoken?"

"That depends, sir," responded Pike. "Have Alastair and Bill been briefed on the OSLO data?"

Churchill observed the negative gestures from both men. "My apologies. I think they can be included."

Menzies and Pike agreed. Sir Geoffrey nodded to Stewart to do the briefing.

"This discussion is also classified MOST SECRET, actually MOST SECRET-OSLO." Menzies paused for an agreement from each man. "In November of last year, our Naval Attaché in Oslo received an anonymous package from an unknown source. The package was a treasure trove of drawings, reports, briefings and test reports purported to be from the very heart of German weapons industry. These data describe turbine engines for aircraft, radio controlled glide bombs, large ballistic rockets, massive artillery guns, radar development, acoustic torpedoes, and the list goes on."

"Has the assessment been completed?" asked Winston.

"Professor Lindemann," began Menzies referring to Churchill's friend and science advisor, "has reviewed the data. R.V. Jones at the Air Ministry is still studying the information. Let it suffice to say, there is sufficient disparity of opinion that we must find corroborative sources to validate each and every item. We know some of it is accurate. We did confirm the flight of the Heinkel One Seven Eight turbine-powered aircraft."

"We have other information of German development work on acoustic fusing," added Pike, "which would also confirm another item."

"Then, the Oslo Report, at least so far, appears to be genuine," stated Churchill.

"I want to caution each of us against premature judgment until each item can be independently verified. Given that caveat, yes, that is certainly my opinion."

"And the atomic development?" probed the First Lord.

"We do not have independent sources as yet," responded 'C.'

"You have been rather quiet, Bill. What do you think?"

Stephenson moved in his chair. "Extraordinary but you know how I feel, Winston. Better to lose a battle than lose a good source of intelligence. The approach presented by Stewart and the fellows is the correct one. I recommend we go with it fully."

"Then, we are agreed," Churchill said as everyone in the room nodded their consent. "Is there anything else we need to discuss?"

"Is it time to inform Roosevelt?" asked Stephenson.

Churchill stared at Stephenson and contemplated his response.

"If you would allow me, First Lord," interjected Pike, "further exposure to ULTRA or OSLO at this juncture would seem to be an unnecessary risk, not worth the potential loss."

"I would agree," added 'C.'

"We should consider . . . ," 'Intrepid' stopped his sentence when Winston held up his right hand.

"Bill, I am as eager as you to engage the President in our little endeavor. We discussed only a month ago Hitler's 29th of November directive for expanding the war. The Phony War will end in the spring. We all know it. Our ULTRA intercept is all we have at this point which leaves it automatically open to question. As much as I would like to engage the President, I am afraid I must agree with 'Jumper' and 'C' at this stage. We need some corroborating information as to Hitler's plan. It would be at that point, when we present our case for the widening war, we should include the President in our little club."

"As you wish."

"First Lord," said Menzies, "have you been briefed on the Case Yellow documents?"

"No."

"Seven days ago, *Luftwaffe* Major Helmut Rhineberger, thought he would get some flight time in a borrowed One Oh Nine as transportation to a staff meeting. Hapless Major Rhineberger did not check the weather forecast, found himself in rather bad weather – fog covering the ground and multi-layered clouds all about – and ran out of fuel trying to find a place to land. Unfortunately for him, his aeroplane made a rather rough landing in farm field in Western Holland." A small rumble of chuckles punctuated Menzies briefing. "The interrogation of Major Rhineberger continues. However, of far greater significance, the major was carrying staff plans and orders for Case Yellow – the German invasion of the Low Countries."

"If they are authentic, that would confirm the decrypt of November 29th, would it not?" asked Winston.

"It would appear so. I must add, First Lord, we are still evaluating the credibility of this information. It's finding was just a little too bizarre and fortuitous for my liking."

"Misinformation?"

"Perhaps. We shall see. This information was with other elements must withstand the rigors of detailed examination and analysis."

"Do we know a date yet?" asked Stephenson.

"No. All we know so far from the information we have collected is spring to summer. Mid-spring would seem to be the earliest to avoid adverse weather."

"You have found no further encrypted information?" asked Churchill.

"No." answered Denniston. "We were extraordinarily lucky on the November 29th message. Our ability to replicate the key settings is far too spotty. We should anticipate more information once we can breakdown the settings. For the moment, that is the best we have."

Churchill looked to each of the others, and then back to Stewart as if he just remembered something. "Stew, have your blokes completed their assessment of our exposure in this Joyce affair?"

"We managed to collect sufficient information to establish the guilt of Lord Haw-Haw."

"Lord Haw-Haw?" asked Stephenson.

"A tabloid reference, I'm afraid," chuckled Stewart, "to William Brooke Joyce, our erstwhile Nazi sympathizer, an Irish-Catholic amalgam, and now confirmed traitor. I might add, for everyone's benefit, we regrettably have now also confirmed that Joyce with his co-conspirators have in fact compromised our Gray codes."

"Gray codes?" asked 'Intrepid.'

"Diplomatic codes we considered to be, practically speaking, unbreakable, at least in any reasonable amount of time," answered Denniston.

"Hopefully," said Winston, "we have our leak plugged."

"Indeed."

"I might add," interjected Menzies, "'K' and his MI5 blokes are working these threads quite well. They have a rather detailed picture, but they are not ready to move."

Major General Sir Vernon George Waldegrave Kell, KBE, CB, known as "K," created the internal Security Service that became known as MI5 – the counter-intelligence group – and he was first and only Director General of MI5. General Kell and Admiral Cumming were colleagues and simultaneously created MI5 and MI6 respectively, in 1909. MI5 focused on domestic threats to national security, while MI6 concentrated on foreign intelligence collection.

"For the time being, Bill, I should think we must press you into service as my messenger with President Roosevelt."

"No problem."

"Brilliant . . . then, we are concluded."

Chapter 19

> Curiosity is one of the most permanent and
> certain characteristics of a vigorous intellect.
> -- Dr. Samuel Johnson

Friday, 2.February.1940
Air Ministry Experimental Station Drone Hill
Cockburnspath, Borders, Scotland

Pilot Officer Brian 'Hunter' Drummond checked the letter of introduction from his group commander and given to him by his squadron leader. 'Spike' Darling had cautioned him to be patient and careful as he approached the RDF station. It was, after all, a sensitive defense site.

A bitterly cold wind blew in directly off the North Sea subtracting from otherwise absorbing scenery of the treeless, emerald green mantled cliffs, rocky beach and perpetual waves. His topcoat, scarf and gloves did little to lessen the stab of the wind. The tangle of barbed wire, chain link fence and barrier with guard shack defined the boundary of what Brian believed was his destination. The pair of Home Guard soldiers in full combat kit left the relative protection of their small hut with rifles and bayonets at the ready. The scene brought a greater chill to the moment.

"Halt where you are," the guard corporal commanded with Brian still five yards away. "State your purpose."

Brian reached into his coat to retrieve his letter of introduction that caused the two guards to raise and aim their rifles at him. The pilot froze rock solid feeling the urge to urinate.

"Whatever you have in there, draw it out very slowly," said the corporal with a hostile, agitated tone. He continued to scan both sides of the road and the area around the checkpoint.

Pilot Officer Drummond did precisely as he was commanded.

"I have a letter allowing me to visit the Drone Hill facility," he said unfolding his letter and holding it high.

The corporal motioned for his companion to advance and evaluate the letter. The man stayed to the side of the narrow gravel road enabling the corporal to maintain a clear line of fire. Brian questioned the wisdom of pursuing his curiosity. At this moment, with a loaded rifle aimed directly at his chest by a rather unpleasant soldier, any value he might derive from the visit did not seem worth it.

"Looks proper, corporal," the younger soldier said.

"Bring it 'ere." After several questioning glances, an assessment of the authenticity of the letter and a thorough check of his identification papers, the corporal shouldered his rifle leading his partner to do the same and waved Brian forward. As he neared the barrier, the two soldiers saluted Brian which he returned. "Welcome to Drone Hill, sir. Flight Lieutenant Harrington is the officer in charge. You will find 'im in the wooden buildin' between the towers."

"Thank you, corporal."

As he walked the distance toward the building and towers he first saw in his low pass a month earlier, he was impressed with the size of the towers. He estimated they had to be 200 to 300 feet high, bigger than they looked from the air. The operation of the towers was still not obvious as he tried to project Roger Beamish's explanation to the physical evidence before him. The fascination and curiosity began to return as the confrontation at the gate fell behind him.

Brian knocked on the only door then entered the building. The warmth of the interior provided an appreciated thawing. A young, reasonably attractive Women Auxiliary Air Force aircraftman sat behind a simple desk in an austere office doing a variety of paperwork. She stood and saluted Brian.

"May I help you, sir?"

"I am here to see the officer in charge."

"Very well, sir. If you will wait here, I will ask Flight Lieutenant Harrington to come out." She disappeared into a darkened room. The peculiar muffled whine of what sounded like a large electric motor came from behind the only interior door. In a minute, she returned with a non-pilot RAF flight lieutenant who was thin and nearly as tall as Brian.

"Yes, pilot officer. I understand you want to see me."

"Yes, sir." Brian handed him the letter and talked while he read. "My group commander has authorized me to visit an RDF site. I am a Spitfire pilot with Six Oh Nine Squadron at RAF Drem. I thought it would help me do my job better."

"I am not sure how your knowledge of this facility will aid your task, but your letter of introduction does appear to be in order. The group commander's office did call ahead to coordinate, so I suppose there is no objection." He paused to return the letter. "Are you an American?"

"Yes, sir."

"Bloody hell."

"What's wrong, sir?"

"Firstly, I've never met an American. Secondly, this is a secret facility conducting very sensitive operations. We don't get many visitors out here, in fact, none. You are the first. Like it or not, you are a foreigner, and I am rather astounded you are permitted access to this facility."

"I am a fighter pilot expected to defend this country," Brian responded with a twinge of resentment over the implied alienation.

"No offense, Mister . . . Drummond, is it?" Brian nodded. "I am simply not accustomed to having any visitors at this remote site, let alone an American volunteer fighter pilot. Nonetheless, your papers are in order and appear to be genuine, therefore, I certainly have no objection to providing you a tour of our humble facility."

"Thank you, sir."

"Well, then, if you will follow me." Flight Lieutenant Harrington led Brian back outside. "The four large towers you see are the antennae through which specific pulses of radio energy are broadcast out to sea to cover our sector. The radio waves travel at the speed of light outward. If they hit an object, like an aeroplane, the waves are reflected back to this site and received by the same antennae."

Harrington waved his hand and led Brian inside through the outer office into a darkened interior room. As his eyes adjusted to the dim light, the details grew from the darkness. A large panel with several knobs, switches, dials and indicators dominated the longest wall opposite the door. Another woman sat at a console with a small work table in front of her. She wore a headset with a horn microphone resting on her chest.

"Aircraftman Morris is our duty system operator," he whispered probably to avoid distracting the woman. "The radio waves we talked about outside come into this rather voluminous bank of the receiver. The signal is transformed into specific peaks on this display," he said pointing over the operator's shoulder at a circular glass screen with a green line at the bottom. Little spikes appeared to randomly jump up from the bottom line with some persisting among the others. "At the moment, we do not have any targets." Harrington looked to the side wall at a one foot diameter clock. "You may get lucky. We usually get a visit from the *Luftwaffe* around this time of the day. While we wait, let me briefly describe how we translate the signals.

"As a bomber approaches, the return radio signal is analyzed, as I said in this behemoth of a box," he said pointing toward the large, wall sized panel in front of them. "A peak appears on the display. The operator adjusted several controls to obtain the best signal. From this display, Aircraftman Morris can determine the position of the target in terms of direction, range and height. From this point, the information is passed via telephone lines to the Filter Room at Fighter Command."

"That's amazing," Brian said.

"Yes, it is, isn't it."

"How do these indications, or whatever you call them, get transformed into commands to us in the cockpit?"

"To be honest, I am not absolutely certain, but I would imagine our detections are combined with other detections from other RDF sites and the Observer Corps. They must do something with the information at the Filter Room."

"Sometimes the controllers tell us how many aircraft are in the raid. How do you determine the number of aircraft?"

"The more aeroplanes, the taller and broader the peak on the display. The operator must be rather well trained and experienced to interpret the peaks that appear, but our Aircraftman Morris is the best."

"Thank you, sir," the woman spoke for the first time.

"Sure doesn't seem like you have many people to run such an important facility?"

"It doesn't take that many, but there are more people than you might think. We need quite a few operators. Working at the display is tiring, and we have been at full alert, twenty-four hour operation, since before the war began. We also have repair and security personnel as well."

"Oh yes," Brian grimaced. "I met some of them upon arrival."

Flight Lieutenant Harrington and Aircraftman Morris both chuckled. "They can be a bit foreboding, but this is a sensitive facility after all."

A change on the display attracted the operator's attention. She began to move several switches and knobs as well as measure a modest size peak now obvious on the right side of the circular screen.

"Ah, here we go. You seem to be in luck, Mister Drummond."

The two men watched and listened as Aircraftman Morris went to work. The process took several minutes before she was satisfied with her measurements. She pushed another button.

"Drone Hill, here," Morris spoke into the horn on her chest. Presumably, she was talking to the so called Filter Room at Fighter Command. "Bogey, estimated to be less than ten aircraft, direction oh eight five, range one one two, height one five."

Harrington leaned toward Brian to whisper again. "By the width and height of the peak, she has determined the size of the raid, less than ten, not big. She has also established the target location at 15,000 feet, oh eight five degrees and one one two miles from here."

"Incredible."

"Yes, it is. Many think this capability is the key to defending our islands."

Morris continued to work with precision and focus reporting the inbound raid. Calls from Drone Hill picked up a rhythm. The reports about every few minutes provided a positive indication of the closing raid.

"Splendid. Good show. You see the new peak on the left side?" Brian nodded. "Those are our fighters climbing to engage." The pace of Morris' activity picked up. Another small target appeared farther out. She calmly, clearly and without a hint of emotion reported her translation of the developing scene. Within about ten minutes, the inbound and outbound peaks merged. "It would appear our boys have bounced the bad guys." The new peak grew to twice the size of the separate peaks, continued inbound for a few minutes, then stopped as if it was frozen in place. The peak gyrated back and forth, growing and contracting, like a strange dance on the screen until it began to move away from the station. In time, the peaks separated again. "See here?" Harrington said pointing to the peak now outbound.

"Yes," Brian whispered.

"Looks like your mates had some good hunting. Nazis appear to be short two or three bombers from the number they started with. Even the other small group further out decided to turn around. Good show. Good show, indeed."

Brian's mind grounded away on the deluge of information coming in. The interception process he had experienced began to make more sense although large gaps in his understanding remained quite apparent. He still did not understand how these radio signals became commands to him in his Spitfire. While Brian did understand the description provided, the details derived by Aircraftman Morris eluded him. So much did not come to him.

"How can you tell our fighters from their bombers, or worse yet, when it is our fighters and their fighters?"

"Without getting into unacceptable areas, let me ask you. Are you familiar with Pipsqueak?"

"Sure. Sometimes the controllers ask us to make sure our Pipsqueak gear is on, or to turn it off and on."

"Yes, well, we detect the signals from your Pipsqueak. It clearly identifies your aircraft from the bad guys who do not have the same emissions. See here," Harrington said as he pointed toward a full scale, spike on the screen, "this is a Pipsqueak transmission. Friendly."

"Brilliant. I never did exactly understand that part. Absolutely incredible."

"Yes, it is."

Brian noticed the clock. He had been at the RDF site for several hours.

It seemed like such a short time. "I'd better be going."

Harrington led Brian out of the operations room. Brian conveyed his gratitude for the tour and description. His level of understanding had grown substantially. He felt better.

"I must remind you, Pilot Officer Drummond . . . what you have seen and learned here is most sensitive, and as you can appreciate, of utmost importance to you and to this country. You cannot be permitted to discuss what you have learned with anyone outside the RAF. I suggest you tuck it away in the back of your brain, use it to perform your part of our defense better, and protect the information. Is that clear?"

"Very clear, sir. I do appreciate the sensitivity. I will do my part."

"Good luck and good hunting."

"Thank you, sir. We'll do our best."

―

Wednesday, 7.February.1940
Headquarters, Secret Intelligence Service
No.21 Queen Anne's Gate
Westminster, London, England

GROUP Captain John Spencer walked a proper, half step behind and to the left of Air Commodore James Hogan, as they entered the non-descript white, heavy oak door with a gold numeral 21 on it. The Headquarters of the Secret Intelligence Service sat near the fringe of the Whitehall district of buildings housing His Majesty's Government. There were no signs, markers, symbols or other indicators that might even remotely offer a passer-by a hint of what lay inside. The two senior air force officers passed patiently through an impressive series of security checks, doors and assessments. Satisfied the two men were, first, who they said they were, and second, who had been specifically invited and so authorized to visit the secret facility, a plain, pale wisp of a man ushered them into a briefing room.

It was not a large room. Small and large scale maps covered most of the walls. A variety of projection equipment occupied the center of the left wall while a high reflectivity screen dominated the opposite wall. A large, sturdy, dark stained, oak conference table along with a full complement of chairs filled the space.

The two officers stood quietly, casually scanning the maps with geography and names they were all too familiar with until the door opened and three men entered. The small statured, dark complexioned, balding, black haired and mustached man introduced himself as Stewart Menzies. John Spencer noted the lack of any pretense from the man who was actually Colonel

Stewart Graham Menzies, DSO, MC, known as 'C,' the Director General of the Secret Intelligence Service or MI6. The other two men were introduced as senior men within his organization.

Carl Acton, Head of Operations, had the appearance of an average businessman or banker of which there were thousands working throughout the city. There were no uniquely distinguishing features to the man including his voice and hand shake. The other man, introduced as Alastair Denniston, was equally non-descript although a little more plain and mousy. Menzies stated his job as 'associated with communications,' which probably meant his position was extremely sensitive. John Spencer wondered if those were their real names.

"Sorry to trouble you with a visit to our humble building, gentlemen," began Menzies, "but, we thought it important to discuss this information here as will become readily apparent in due course. Before my able assistants begin, I must strongly caution you, the information you will soon discuss is MOST SECRET and cannot be discussed outside this room. Mister Acton will address the issue of what can be discussed in a properly controlled environment and only with individuals with a critical need-to-know. Any questions?" Both air force officers shook their heads. "Excellent. Then, I will leave you to get on with this." Menzies did not wait for an acknowledgment.

Acton began the presentation. "We understand that both of you are essentially aware of the turbine engine development undertaken by Wing Commander Frank Whittle. Correct?" Both men nodded. "Excellent. It is also our understanding you are familiar with comparable work going on in Germany."

Air Commodore Hogan spoke for the air force. "We have seen several elements of information, the latest being October last, I believe. Familiar might be a stretch for us."

"Fine. As you can obviously surmise, we want to assess some recent data regarding the German turbine engine development program and its application to fighter aeroplanes."

"All right," responded Hogan without a hint of the curiosity John Spencer felt.

"We have indications the German development of jet fighters is progressing quite rapidly."

John knew the consequences of that simple statement, and it gave him a chill. He was quite familiar with Frank Whittle's calculations regarding the possibilities of turbine-powered aircraft. The speed potential, lacking the drag of a conventional propeller, significantly exceeded any current fighter on either side. He was also intimately aware and appreciative of the fighter pilot's axiom from his combat days – speed is life. There was an essential, inviolate truth to the phrase.

"The two principal camps are Heinkel with a single turbine engine, aluminium, monoplane designated, Model One Seven Eight, and Messerschmitt with a twin engine version designated, Model Two Six Two. Willie Messerschmitt's aircraft seems to be having more problems primarily because it is somewhat more sophisticated requiring more thrust from a smaller diameter engine. His machine apparently has the engines mounted either in or under the wing separate from the fuselage while the Heinkel machine has as single engine, as I said earlier, embedded within the fuselage. The Heinkel One Seven Eight flew its first flight last August." The statement hit John Spencer like a lightning bolt. "The best information we have is, the Messerschmitt Two Six Two may be as much as two years behind."

"Thank God for that," John said without thinking.

"Yes, quite," Acton continued. "We also believe Heinkel is having some difficulty with the engine technology. From what we know of Whittle's work, it would seem we are about one to two years behind the Germans."

"Agreed," said John without being asked.

"Given this information, when do you think the Germans might have an operational jet fighter?"

The question actually staggered Group Captain Spencer. The reality of a terrible nightmare never made anyone feel better. The prospect of RAF fighter pilots, even with airplanes as impressive as the Spitfire and Hurricane, facing jet engined adversaries with a hundred miles an hour or more speed advantage knotted his stomach, burning his insides. The sensation immediately reminded him so vividly of the feeling he got 23 years ago when he had to face the quite new Fokker D.7 Tri-winged, German fighter with his older Sopwith 1½ Strutter before they received the more capable Sopwith F.1 Camel. It was not a good feeling. His mind briefly flashed to his best friend's young protégé in a Spitfire fighter squadron near Edinburgh as well as other young pilots he knew. A terrible darkening to his mood brought an even worse burning to his gut.

"Given the first flight of the Heinkel machine last summer, the normal aeroplane development time adjusted for the usual delays associated with advanced equipment, they could have an operational jet fighter by this coming summer." John Spencer knew his words were correct although he did not want to believe them.

"Summer, aye?"

"Yes."

"Group Captain Spencer, I do believe that is a Distinguished Flying Cross on your tunic." John nodded his confirmation. "As an accomplished fighter pilot yourself, what does the advent of adversary turbine fighters mean to you?"

"I was considering that very question a short time ago. Let me, if I can, put it simply. If Whittle's calculations are correct, a jet fighter will have as much as a hundred mile an hour speed advantage over our best fighter, the Spitfire. That sort of difference, and I will try not to overstate it, is like trying to fight with a club against an opponent using a rifle."

"That bad?" asked Acton.

"That bad!"

"Well, then, I should say our work is cut out for us," Acton interjected.

John considered the meaning of Carl Acton's words as the head of SIS Operations – most likely, a new focus target for not only intelligence collection, but presumably some sort of offensive effort to slow down or stop the German aviation development – a great objective from John's perspective, but beyond his experience and knowledge. For the first time, John Spencer wondered what the other SIS man, Denniston, actually did within the service. Menzies said communications, but maybe he was actually an agent. He certainly did not look like John's image of a spook, a field agent.

"A few more bits may be of interest to you. We also have indications of a new fighter called, a Folke-Wulf One Nine Oh. The extent of our knowledge is, it is a single seat, radial engine, aluminium, monoplane fighter. References to the One Nine Oh lead us to believe it may be faster and more maneuverable than the current production Messerschmitt One Oh Nine."

"Oh, great. Simply great," interjected Group Captain Spencer. "How much more good news do you blokes have?"

"John," cautioned Air Commodore Hogan. "These gentlemen are trying to help us by making sure we have the best possible information about our enemy. Let's not strap the messenger."

"Quite right. So sorry. My apologies. I simply have a great deal of empathy for our young fighter pilots and what lies before them."

"Understood, Group Captain Spencer. Then, it might be of some comfort that the Germans have apparently significantly slowed the development and production of the Folke-Wulf One Nine Oh in favor of increased production of the Messerschmitt One Oh Nine."

"Not much comfort, I must say. The One Oh Nine is a very capable fighter, roughly the equal of the Spitfire. In the hands of an experienced and capable pilot, it is a formidable opponent."

"Understood. The highest levels of the *Luftwaffe*, including the air fleet commanders although we do not believe below that level, have been informed of various aspects of fighter development by German industry. This suggests that the operational commanders do not share the enthusiasm of the engineers. They apparently want a proven bird in hand. They have experience and faith in the One Oh Nine and, it would seem, are not particularly impressed with the development work. Maybe there is some brightness in this rather gloomy picture."

Both RAF officers nodded. John thought Hogan was probably absorbing the information and its implications as he was. There was not much to be hopeful about in the briefing.

"I'm afraid that is the extent of the information we have so far. Any questions?"

"Probably more questions than there are answers, but none at the moment," answered Air Commodore Hogan. "I suspect my colleague feels as I do. We must digest what we have heard."

"Yes, quite. Well, then, I must reinforce Director General Menzies's admonition. This information is MOST SECRET and must not be discussed with inappropriate individuals." The recipients nodded. "We will produce a scrubbed version of this information along with our analysis as you have suggested. It should be available to the Air Ministry tomorrow. We shall also do what we can to learn more, try to affect their development efforts and encourage the highest priority support for Wing Commander Whittle's development program."

"We could not ask for more," answered Air Commodore Hogan.

"Thank you. We will contact you when we have more information, or need your assistance."

"Thank you, and good luck."

They left the SIS building to return to Bentley Priory. Neither man said a word during the journey out of town to Fighter Command headquarters north of London. Although it happened to be a bright, but cold day, it was a dark, very dark, day for Group Captain Spencer. The burden of what lay ahead approached the level of unbearable, but he also knew the only way to weather the storm was to batten down, hold on and make it to the other side. He felt empty, nonetheless.

Thursday, 8.February.1940
RAF Edinburgh
Edinburgh, Lothian, Scotland

The grips of what many called the worst winter in ages held tight the space outside the pilot's barracks. Green Section, No.609 Squadron, had this day for rest. Pilot Officer Brian Drummond had come to appreciate the wisdom of the liberal rest and relaxation holiday policy of Fighter Command. Although the flights were until now relatively short and infrequent, the tension and pressure of fighter operations possessed a uniquely fatiguing quality. Some outside the community considered the policy too liberal, but no one inside did. On this day, part of Brian's consciousness wanted to remain inside to read, nap and write letters, but he had asked for and been granted access to the Turnhouse Sector Control building at RAF Edinburgh. While it would be easy to give into the desire to stay warm, he knew he had to go for a variety of reasons. A number of special arrangements had been made despite resistance from several places. People were expecting him now that permission had been granted.

Brian hitched a ride in the twice daily lorry making the journey across town to RAF Edinburgh. The heater in the lorry barely kept the cold out, but Brian was thankful it was warmer than his Spitfire cockpit at 20,000 feet. The journey took several hours although the distance was not particularly far. Numerous roadblock checkpoints, normal morning traffic and the requirement to pass through the center of Edinburgh from the Eastside added the majority of time.

The aerodrome sentries, although just as unfriendly as those he encountered at AMES Drone Hill, were thankfully less threatening. Although he had flown into RAF Edinburgh several times, Brian had never entered the base from the roadway. Buildings of brick, concrete and steel added a sense of permanence that RAF Drem did not enjoy. Following the directions of the gate guards, the driver dropped him at a building whose corrugated tin roof was the only element visible. It was surrounded by a high, thick, dirt berm backed by a concrete wall. Grass covered the slope not blanketed by accumulated snow. Only one entrance through the fortification was evident. A small, low sign identified the place – Sector Control. Two burly, alert and menacing guards stood on either side of the entrance.

As Brian climbed down from the cab of the lorry, the two guards watched disapprovingly. After examining his RAF identification card and his letter of permission for access, the senior guard directed him to pass into the area, turn right and enter through the center door. He was to ask for Flight

Lieutenant Lomstrong, the duty officer. As he passed through the zigzag, concrete lined entryway, Brian was surprised to find a green, conventional clapboard, military building. Entering the center door as instructed, he stood in a narrow hallway with doors on either end and three more doors along the corridor. An open office door greeted him. An aircraftman sat behind the desk doing paperwork. When asked, he rose to retrieve the requested officer.

"You must be Pilot Officer Drummond from Six Oh Nine Squadron," stated the flight lieutenant emerging from the far right door. He did not wear pilot's wings nor any decorations.

"Yes, sir."

"Excellent. Name's Lomstrong. I am the current duty officer. I tend to the administrative stuff while the controller runs the show. May I be so bold . . . did I detect a Canadian accent?"

"No, sir. American."

"Bloody hell," he responded with astonishment. "Yank, aye. Didn't know we had any Yanks in the RAF."

"As far as I know, I'm the first and maybe only one so far. I'll bet there will be others, if they aren't already here."

"Bless you, then, nonetheless. We certainly do need all the fighter pilots we can get at the moment, unfortunately. So . . . as I understand it, you wanted to see how sector control works. I imagine it would seem a bit mysterious being in a cockpit not knowing how you are vectored to the target. Before we begin, I am afraid I must see your authorization letter."

Brian produced his letter which satisfied Flight Lieutenant Lomstrong. His host described the facility as devoted to the conduct of fighter operations. The other spaces existed for only one purpose – support the sector control room.

Lomstrong stopped before he opened the far right door, the door he first appeared from. "Before we go into the control room, I am afraid we must be most discreet with our words. Operations are underway and disturbance of the process is not tolerated."

They entered a room bursting with detail. Brian carefully absorbed as much as he possibly could without any words from Flight Lieutenant Lomstrong. The entry door opened onto a narrow sub-balcony passageway to another door. The highest portion of the room along the left wall entailed a series of desk-like work stations. A wing commander occupied the middle desk overlooking the entire room. He was the sector controller, if Brian guessed correctly. Closest to the door were several communications stations manned by two squadron leaders with several telephones and radios each. An enormous map table depicting Eastern Scotland, the Northeastern portion of England and the North Sea approaches occupied the main part of the room, about four feet below the sub-balcony upon which Brian and Lomstrong stood and seven feet below the controller's desk. A grid with each square annotated by unique code letters divided up the entire map. A half dozen female aircraftmen surrounded the three sides away from the controller, so he had an unobstructed view of the entire table. Each aircraftman had a headset and a horned phone like the operator at Drone Hill and was armed with a long wooden hoe like a croupier at a gaming table. Along the wall opposite the controller, a series of multi-colored tote boards defined the operational status of each squadron under Turnhouse Sector Control. Brian noted his squadron identified by lights as being available within 30 minutes. Several squadrons, or elements of squadrons, were airborne across the board.

A jumble of various radio communications broadcast from a loudspeaker offered the only audio other than muffled words of various people in the room and commands from the senior controller. The radio calls were the familiar jargon and chatter of fighter pilots and their controllers. From what Brian could ascertain, there were at least three interceptions in action in the sector.

Brian intently observed the activity on the table associating commands by the senior controller as well as the communicators with movements of objects on the map table. He quickly figured out the large blocks being moved across the table were enemy raids, each noted by an identification number prefaced with the letter 'h' with the number of enemy aircraft and height below the ID number. A different block annotated by a sortie number prefaced with the letter 'F' and altitude denoted a friendly squadron. Each friendly block had a little flag with the appropriate squadron affixed to it.

The WAAF coordinators had to be in contact with the Filter Room at Fighter Command, maybe through the group headquarters, since that is where Flight Lieutenant Harrington at Drone Hill Chain Home Station said the RDF detection information went for collation with other indications. As he watched the action, he told himself he had to contact Group Captain Spencer for a visit to Fighter Command.

Brian leaned toward Flight Lieutenant Lomstrong and whispered, "How do they determine where to position the fighters?"

"You see the blocks on the table?"

"I've figured all those out. How do they position a squadron to intercept an incoming raid?"

"Right, then. The controllers, these gentlemen sitting over here," he said pointing to the flight lieutenants Brian thought were communicators, "must judge the direction and rate of progress of an inbound raid. Once the senior controller commits an element to an intercept, the controller judges the progress of the fighters. They use a technique called, Tizzy's Angle, named after Sir Henry Tizard, Scientific Advisor to the Chief of the Air Staff, which is a specific application of the principle of equal angles. The controller wants to create a condition where the bearing to the target remains the same and the range to the target decreases. Other factors the controllers use are certainly the time of day for sun angle, cloud cover for masking the approach and the closeness to landfall."

"Would you, please, explain this Tizzy's Angle again, sir?"

"Here, let me try it this way." Lomstrong pointed to the map table. "Watch the raid identified as 'h' Eight One Three. It says it is a hostile flight of less than five aircraft, probably a solo reconnaissance bomber. It is at 18,000 feet. Now, watch the direction and rate of travel for 'h' Eight One Three. Do you see 'F' One Seven?"

"Yes, sir."

"As you can see, it is a sortie by Seven Two Squadron out of Acklington, probably one section of three Spitfires. Now, watch the direction and rate of travel of 'F' One Seven." Lomstrong paused for several minutes to let Brian evaluate the movement of the two blocks. "Do you see the constant bearing decreasing range between the two?"

"By God," Brian said with excitement.

"Quiet," snapped the senior controller.

"Mister Drummond, you will have to contain your enthusiasm, or we will be evicted."

"Yes, sir. Sorry, sir."

"No problem, actually. You are not the first to react to this particular revelation like that." Lomstrong gave Brian a pat on the back. "Now, other factors. The clock says nearly twelve noon. The sun is at its zenith, but since it is winter, it is relatively low on the horizon. If you will further note, the controller selected a squadron to the South of the inbound track enabling the fighters to intercept from the South, out of the sun."

"I'll be damned."

"I hope not. We don't seem to have any fighter pilots to spare these days," Lomstrong said with a chuckle. "It looks as if our 'h' Eight One Three intruders decided they did not want to tangle with Spitfires. Unfortunately for them, the Spitfires are within striking distance."

They both watched. Brian remembered watching the two peaks on the

RDF screen at Drone Hill merge, gyrate and then separate. The fascination of this aspect of the air defense problem absorbed Brian's concentration. Learning how the system worked and watching the performance helped him understand and appreciate his limited experience as well as the stories of the more experienced pilots with fighter intercepts. He now realized to a greater extent than ever before the process was not magic or happenstance.

Within minutes, one of the WAAF coordinators reached out with her hoe and racked off the 'h' Eight One Three block. A short time later, the 'F' One Seven block began moving southwest toward RAF Acklington, north of Newcastle.

"One more victory for the good guys. They can run, but they cannot hide," Lomstrong chuckled quietly.

While there was probably much more he could learn, Brian had what he needed, a view of how the mechanics of an intercept was set up. "My transport back to Drem should be about ready to leave. I'd better go. Thank you for your help, sir. You have been most helpful."

"Do good work, Mister Drummond. We'll do our part, but we're counting on you and your mates."

The ride back to RAF Drem passed quickly with Brian lost in his thoughts about the ADGB, the Air Defense system of Great Britain. Brian waited in the pilot's barracks until near sunset when Jonathan returned. He had two sorties. One occurred before and the other after Brian's visit to Turnhouse Sector Control while he was enroute. Jonathan looked tired. Two sorties and no engagements. Brian wondered as Jonathan did when they would actually fire their guns at a real target. They both could only imagine what the experience would be like.

After listening to Jonathan's frustration with no engagements, Brian conveyed his excitement of learning. They bantered through the evening meal and into the night about how to use Brian's new found knowledge. While it probably would not help them fly their aircraft better or shoot more precisely, it did help them to understand the complexity of the interception problem. Brian was feeling better, but he still needed to visit Fighter Command. He told Jonathan about his intention to ask Group Captain Spencer for assistance.

"Do you want to go with me to Stanmore?" asked Brian.

"Sure. While we are in London, we can see Linda and Anne."

"Sounds like a plan."

Friday, 9.February.1940
Headquarters, Fighter Command
Bentley Priory
Stanmore, Middlesex, England

"Group Captain Spencer," he said automatically into the telephone.

"Sir, this is the Filter Room. It looks like we bagged another Heinkel One Eleven. It was a solo. Our guess was the bloody bastard was on a photographic reconnaissance mission."

"I assume since you are calling me, the aircraft survived the landing."

"Yes, sir. The fighter boys followed it down making several low passes. According to the initial reports. The bomber crew did a rather good job getting their wounded bird on the ground with only minimal damage. They landed in an open field near an anti-aircraft gun battery. The bomber crew was captured before they could destroy anything."

"The good Lord does bless us now and again. Anything else?"

"Not at present. We will keep you informed as we learn more."

"Brilliant. Oh, yes, one more thing. Where is it?"

"Near Newcastle."

John Spencer immediately thought of Pilot Officer Brian Drummond. "Who brought her down?"

"Six Oh Nine out of Drem."

A clearer image of Brian blossomed in his thoughts. He would ask Brian the next time he saw him. "Thank you," John said, then depressed the telephone action, released it and dialed an extension.

"Intell," the man said.

"This is Group Captain Spencer. Air Commodore Hogan, please"

The transfer was prompt as usual. "Yes, John."

"Appears we may have another specimen, air commodore. Heinkel One One One near Newcastle."

"Excellent. I shall gather up the things we need."

"I will be right there," Spencer said and hung up the telephone. He walked down the ground floor hallway to the stairs leading to the operations bunker some forty feet underground.

Fighter Command's principal operations complex occupied the majority of the subterranean bunker. Several other critical functions used peripheral rooms in the underground facility including an essential command and control capability, and the Intelligence Section. The plain, unadorned, concrete walls made the underground complex seem so much colder and sterile. John Spencer never did like the place. He could never establish whether it was the décor of the bunker, or the fact there were no windows. He could not see the sky.

The Intelligence Section occupied a corner office area behind a combination locked, steel door. John Spencer always wondered why someone felt the need to secure the intelligence function in this way forty feet below the Fighter Command headquarters building, one of the most secure RAF facilities. He rang the entry buzzer.

The small slot in the door opened revealing a familiar pair of eyes. The slot closed and the door opened. "Good afternoon, sergeant."

"Good afternoon to you, sir. The air commodore is in his office."

"Thank you, sergeant."

John worked his way past the filing cabinets containing the seemingly endless catalogue of information about their enemy and even some friends, past the light tables for photographic assessment, and past a wide variety of maps. Air Commodore Hogan resided in the corner office of the corner set of offices.

"I've made a few phone calls. The group intelligence officer moved a security and exploitation team to the site in short order. He has confirmed the downed bomber is in good shape. It does not contain the *Knickebein* equipment we found in the October capture."

"Does he think the aircraft can be made flyable?"

"Yes. He says there is damage to the undercarriage, airscrews and some skin damage, but there is no other obvious damage."

"Good." John thought about a series of flight trials he might organize to evaluate potential weak points in the bomber's defense systems. He did not waste too much thought on the notional exercise since that was the job of the professionals at the Royal Aeronautical Establishment at Farnborough. They were well-qualified, experienced and capable of performing the exploitation task. "Have the lads at Farnborough been notified?"

"If they have not, they will be soon."

"What did they do with the *Knickebein* gadget we captured in October?"

"As you may remember, the One Eleven we captured with the equipment turned out to be more damaged than initially thought. The *Knickebein* items were documented, removed and installed in a Hudson test aeroplane. We have had very good results. We have definitively established the purpose, operation and capabilities of the system."

"So, it is a foul weather bombing system?"

"That and more."

"Such as?"

"In addition to reasonably accurate night and foul weather bombing, it offers a night flight routing and recovery capability as well."

The weight of German aviation development and ingenuity redoubled as he considered the growing list of enemy capabilities the RAF did not possess. The experts tried in earnest to learn as much as they could and to exploit it, meaning to either find ways of defeating it or to use it on our side. Business for the engineers and pilots at RAE Farnborough was picking up. One of many recurring questions in Group Captain Spencer's mind was, would they learn enough in time to meet the growing inevitability of the onslaught?

"Somehow we must slow down the aviation development in Germany," John suggested almost to himself.

"There are those who are working on just that, but we do seem to have a rather current problem ourselves."

"Yes, indeed. Well . . . I think we should give the blokes at RAE a call. They need to be on this one as quickly as possible. If that bomber is repairable, we need to get our fighter pilots in and around it. They need to appreciate its capabilities and limitations."

"Certainly, John. We will keep you posted as usual."

"Thank you, commodore. Cheers," John Spencer said as he departed. He was lost in his thoughts as he returned to his above-ground office. Unfortunately for him, his concentration and detachment nearly caused him to collide with his boss, Air Chief Marshal Dowding.

"A bit distracted there, John," Dowding said in a distant, yet disapproving manner.

"Yes, sir. My apologies, sir."

"What is it this time?"

"We just brought down another bomber, a Heinkel One One One, in rather good shape, it appears."

"That is fairly good news, I should think. Why should that be so troubling?"

"In itself, no, sir. It is just that Air Commodore Hogan and I participated in a rather troubling discussion a few days ago, and these latest intrusions coupled with the weight of German aviation development is a bit daunting."

John Spencer was rewarded with a rare smile from Dowding. "Yes, well, I am sure you will recall what some of us believed to be the clarion call from your uncle quite a few years back. We certainly do not gain much comfort from knowing he was precisely correct back then, now do we?"

"No, sir."

"Brilliant. If you would allow me, I might suggest you throttle back just a smidgen before you knock some unsuspecting WAAF on her bum."

"Yes, sir," John responded with a chuckle. Dowding moved on down the hall. John turned to watch his commander. He did not always understand why he was given the nickname, 'Stuffy.' He saw a side of Dowding maybe most people did not. Among the fighter pilot community, he was revered virtually as a saint much as he and his brethren respected 'Boom' Trenchard. He had done more to improve the performance and survival of a fighter pilot in the RAF than almost any other single individual in the world. The facet John appreciated the most was his empathy. The man cared, more than most people would ever know. The pilots jokingly referred to themselves as Dowding's 'chicks,' and it was said with great affection. While Group Captain Spencer felt the burden of the boiling storm on the horizon, Dowding bore the full weight in all its elements, facets and dimensions. Sometimes he wondered, how Dowding stood the pressure.

"Sir, a Pilot Officer Drummond called," announced his clerk as he entered his office, two doors down and across the hallway from Dowding's. "I told him you were unavailable. He would only say his call was personal."

What could be the problem with Brian? Maybe he was having second thoughts about being a volunteer with aerial combat coming closer by the day? Maybe that American embassy man was back despite efforts to avoid any confrontation? The RAF, and especially Fighter Command, could not afford to lose even one pilot.

"Six Oh Nine Operations," the female voice stated.

"This is Group Captain Spencer, calling for Pilot Officer Drummond."

"Yes, sir. One moment please, sir," she responded.

John could hear the confident feminine voice call Brian even though she had covered the mouth piece. He smiled as he also heard the hoots and jabs at Brian for receiving a call from a senior officer while he was on duty.

"Thank you for calling, sir."

Hearing Brian's cheerful, strong and robust voice made John take a deep, clearing breath. The humor, laughter and jokes reminded him of his own squadron in the Great War when they had reason to celebrate a victory. Maybe there were bright spots in an otherwise dark and cloudy world. "Good to hear from you, Brian. What can I do for you?"

"Well, sir, I know you are very busy, and I apologize for bothering you."

"No bother," he said not being entirely truthful.

"Sir, I was recently able to visit a Chain Home site and just yesterday visited our sector control station. I've been trying to understand how our fighters fit within the air defense system. Besides my curiosity about how the process works, I hope to be a better fighter pilot if I understand how my aircraft functions as the pointy end of the sword."

John laughed deeply forcing him to recognize it had been sometime since he had been able to laugh so easily. "Yes. You are definitely the pointy end of the sword, as you say. It is good you were able to visit those important facilities," he said with a thought to himself about security. Did a pilot really need to know how the system worked? If Brian where, God forbid, shot down and captured, would he be an unacceptable exposure to the United Kingdom's air defense network? Maybe it was a risk worth taking, if it made him and others more effective fighter pilots? It was refreshing to John knowing the young man on the other end of the phone exhibited many of the same attributes so evident in his best friend, Malcolm Bainbridge. Maybe they could withstand the German invasion with pilots like Brian, like Malcolm. "I suppose your next step is to visit Fighter Command. You've seen the spokes. Now, you want to see the hub."

"Why, yes, sir, that is exactly what I wanted to know?" Brian said sounding a bit surprised.

John laughed again. It felt so good to laugh. The lightning of youth coursed through his body from the other end of the telephone. To his regret, reality returned. "Brian, has someone explained the sensitivity of the places you have visited?"

"Yes, sir, in a very personal way. I am well aware of the importance of each facility. I don't want to cause any trouble. I just want to be a better fighter pilot."

"Well-stated, my boy. I must tell you we don't usually conduct tours and when we do, it is for MPs, senior officers or other dignitaries."

"I understand, sir," Brian jumped in with clear disappointment in his words. "I just thought I would ask."

"Not so fast, my boy. While it is rather extraordinary to have an operational fighter pilot visit this place, you and your mates are a pretty damn important element in this system. If a visit helps you blokes kill one more German, it is worth the risk. Let me check around and prepare the ground a little before I confirm permission for you to visit. I cannot guarantee I will be successful, but I will certainly try.

"Thank you, sir. I really appreciate it."

"By the way, Brian. We understand your squadron brought a bomber down."

"Yes, sir. Our skipper, Squadron Leader 'Spike' Darling got his second kill, a Heinkel One Eleven. Managed to land without much damage from what the skipper says."

"Good show. Please pass my congratulations to your squadron leader." The burden of what he knew enveloped him again. "Good hunting, Brian. Keep your eyes moving, always. Always, Brian! Do not relax an instant when you are in the air."

"Yes, sir. We'll do our best."

"I am certain you will."

"Thank you, sir."

"I'll try to get things sorted out here. I'll be back to you either way."

"I look forward to it. Cheers, sir."

John smiled hearing Brian's use of a uniquely British idiom. "Cheers, Brian." John Spencer returned the handset to the cradle and placed his head in his hands. My God, why is this happening, he said to his most brutal of memories.

―

CHAPTER 20

> When peace has been broken anywhere,
> the peace of all countries everywhere is in danger.
> -- Franklin Delano Roosevelt
> 3.September.1939

Thursday, 15.February.1940
RAF Drem
Drem, Lothian, Scotland

ALTHOUGH all pilots tried to ignore the waiting and each of them had their methods to deal with the pressure and tension of waiting, the hours never became easier. Fortunately, the controllers recognized the mental fatigue of pilots standing ready to launch into unknown situations against a determined adversary. Usually, they were only left in available status for a half day. This day was different. The entire squadron had been on 30 minute alert since sunrise. They had gone to readiness once in the morning and all the way to standby twice in the afternoon. Every time it happened, frustration invariably stretched their tolerance and patience.

Squadron Leader Darling returned to Dispersal from his usual afternoon walk assessing the state of his squadron. The smile on his face meant good news. "Well, lads, it's finally our turn."

"What's that skipper?" asked Flight Lieutenant Beamish.

"We are next for the Mod IA retrofit."

The cheers of appreciation mixed with other jeers deriding the delays brought a strange circus atmosphere to the small building. Every Spitfire pilot in the RAF knew the extent of the Mod IA changes. The extra 200 horsepower from the new Merlin III engine combined with the new 100 octane fuel and especially the three-bladed, metal, Rotol propeller with its DeHavilland constant speed unit meant another 45 miles an hour top end speed. The reflex gun sight with stadiametric range adjustment along with the incorporation of a gun camera gave the fighter pilots better engagement capability. The film would document target results as well as provide a clear learning tool. The addition of armor plating under and behind the pilot's seat as well as the bullet proof glass directly in front of the new gun sight added protection. The hydraulic undercarriage modification eliminated the hand pump and the 'Spitfire knuckles' skinned trying to raise the landing gear. The changes probably had more value in the added confidence each pilot received with the enhanced capability of the Vickers Supermarine Spitfire Mark IA fighter – a great fighter became better.

"Who gets the first mod?" asked Pilot Officer Stephan 'Mongo'

Strickland.

"Come now, lads, there must be some advantage to this thankless job," answered Darling. The hoots and hollers added to the levity. "First mods begin next week and should take about six weeks to complete. Here is the schedule. We tried to arrange maintenance time around each pilot's planned holiday time."

The list told Brian his aircraft was the next to last machine which meant he would not get his new fighter until late March or early April. They were all eager to grab ahold of the new capability. All the reports from other squadrons ahead of them had been exceptional. The enthusiasm of Darling's announcement soon gave way to the monotony of alert waiting.

The telephone rang.

"The squadron is released," announced Corporal Warren.

"It's about soddin' time," responded Flying Sergeant 'Fog' Johnson.

The pilots stowed whatever items of their flying kit still with them. As some of the pilots left Dispersal, the mail was delivered somewhat late. Brian had two letters, one from Miss Rebecca Seward and one from Headquarters, Fighter Command. He had not heard from Becky since before Christmas. He suspected what the letter would say. Brian decided to read Becky's letter away from the critical observation of the other pilots.

"I'll catch up with you in a few minutes," Brian said to Jonathan Kensington.

Concern came from his friend's eyes. "Is everything OK?" Jonathan asked.

"Sure, sure. No problem."

"As you say, then. I'll see you before evening meal?"

"Sure."

Only Corporal Warren remained although she appeared to be quite absorbed in completing the usual myriad of reports and other paperwork. There were always reports to be filed even though nothing happened. It seemed to be a tradition in all military organizations no matter what the color of the uniform. Brian picked a chair facing away from the operations clerk.

January 19, 1940
Dear Brian,

    I don't know how to do this any easier, so I just have to jump into it.

    It is obvious from your letters you have no intention of returning home from this foolish adventure of yours. Everyone we know thinks you are crazy for doing what you are doing. Most believe there is no war, and even if there was a war, it is a European war, NOT an American war.

    I saw your parents over the Christmas break. They are still very angry with you, but I suppose you know that. They are taking steps through Senator Capper's office and I think the FBI to make you return to Wichita. I know how you feel about flying your beloved Spitfire, whatever that is, so I know you won't come back voluntarily. I just hope all this is worth it to you, Brian. I loved you so much I gave myself to you and that was very special, but I guess not that special to you. I'm sorry, Brian.

    You should be here in Lawrence with me. The University is such a beautiful place. The people and professors are fabulous. I really enjoy my studies and I know you would have enjoyed it here as well. Anyway, I've met another man, a veterinary student like me, but a year ahead of me. He is nice, Brian, and reminds me you, but you are not here. I don't know how much longer I can resist the temptation.

    I hope you don't do something

> foolish after reading this letter. In many ways I still love you, Brian, but I just can't wait for you to get this out of your system. I am truly sorry. Please be careful.
>
> Becky

Knowing what the letter would probably say did not make the words any easier to read. Brian knew in his heart the letter would probably come eventually but did not expect it so soon. Why is it no one understands me, he asked himself? Why can't anyone see why this is so important? Brian started to crumple the letter then stopped. After smoothing it out and refolding it, he returned it to its envelope and put it in his shirt pocket inside his tunic.

Although Brian did not think of his relationship with either Anne Booth or Rosemary Kensington, he still felt a sense of loss. An important episode in his life seemed to be slipping through his grasp. The image of her face came to him for the first time in several weeks. There was an edge of guilt for what sacrifices were associated with his dream. He wanted to feel her, to give what he had not been able to give her before. She deserved so much more. The tug between his dream and his heart brought questions of value. Brian told himself several times that he was doing the correct thing. The seduction of high energy flight and a sense of history convinced him.

"Are you all right, sir?" asked Corporal Warren in her soft, caring voice. He must have slumped in his chair lost in thought, he guessed.

"Quite all right, actually. Thank you for asking, Corporal Warren."

"Right, then. Would you be so kind to extinguish the lights and close the door when you leave?"

"My pleasure. See you in the morning."

"Good night, sir."

Brian returned to his thoughts. The image of Becky did not come back to him. The other letter. What did Fighter Command want? It looked exactly like the envelope that contained his orders to Spitfires after completing training at OTU7. Oh God, was he being transferred to another squadron or worst being released from the RAF for extradition to America? "Please, no," he said aloud to himself. He stared at the unopened envelope. With rejection by Becky, a rejection by Fighter Command would be consistent and apropos. Maybe he should call Group Captain Spencer before he opened the letter, so he could say he had not seen it and might be able to change the dreadful message.

Brian stood to raise the telephone. He started to dial the number he remembered. The clock said, 18:37. The handset returned to the cradle without completion of the call. He probably was not there anyway, he told himself. The letter loomed in front of him. A deep breath marked his conviction to deal with the contents.

HEADQUARTERS, FIGHTER COMMAND
ROYAL AIR FORCE,
BENTLEY PRIORY, STANMORE, MIDDLESEX

Telephone Nos.:   BUSHEY HEATH 1661 (6 lines)
                   BUSHEY HEATH 1646 (4 lines).

Telegraphic Address: "AIRGENARCH STANMORE."

Reference: -- FC/S.18002
12th February, 1940
To: Pilot Officer Brian A. Drummond
No.609 (West Riding) Squadron
RAF Drem
Drem, Lothian, Scotland

    Commensurate with your duties to the Crown as a fighter pilot and in recognition of your request, you are hereby granted permission to visit Headquarters, Fighter Command, at your convenience. This letter shall serve as your authorisation to enter the Headquarters compound.

    2. Your host will be:
        Group Captain John H.R. Spencer, DFC
        Staff Secretary to Fighter Command.

    3. You are to notify this command of your expected arrival time at least 24 hours prior to facilitate proper security arrangements.

    4. By the distribution below, your superiors have been informed of your visit.

        H.C.T. Dowding
        Air Chief Marshal,
    Air Officer Commanding-in-Chief
    Fighter Command, Royal Air Force.

```
Distribution:
AOC-in-C, No.13 Group
CO, No.609 Squadron
FC Staff Secretary
FC Staff 2-sec
File
```

"Yehaw," Brian shouted with no one to dampen his ardor. The contrast between imagination and reality heightened his excitement.

As he walked briskly toward the pilot's barracks oblivious to the cold night air, he held the letter tightly and wore a broad smile. Several enlisted men saluted and returned a curious expression over the unbridled enthusiasm of the American volunteer fighter pilot.

"I've got it," Brian told Jonathan a little more loudly than necessary.

"Got what?"

"Yes, indeed, 'Hunter.' What do you have?" asked 'Mongo' Strickland.

"I've got permission to visit Fighter Command."

"You know, for a Yank, you seem to be a bloody gentleman of privilege," interjected Flying Officer Reggie 'Organ' Foxworth, the left wing of Yellow Section and named for the scale of a particular part of his anatomy.

"Leave the lad alone," 'Jackstay' jumped to Brian's defense. "You're just jealous."

"Jealous of what?"

"You wish you had a benefactor. Don't hold it against 'Hunter' because he does. It's life. He simply wants to know how things work. Don't blame him because you don't."

"Well, now, aren't we protective of our fledgling," answered 'Organ' with the most sarcasm he could muster.

The verbal volley continued as Jonathan turned to Brian. "When are you thinking of going?"

"My next two-day pass."

"Which is . . . a fortnight?"

"Not quite. Monday week."

Jonathan considered the words. "That doesn't match up with mine, obviously, so, go ahead without me."

Brian wanted his friend to go with him. "Let's ask the skipper if we can make a switch."

"You know how he feels about that."

"Yeah, sure, but this is an exception," Brian said. "It never hurts to ask," he added automatically.

"Yes, it does depending on what mood he is in."

"If we go down, we can meet Linda and Anne."

"All right. As you say, what can it hurt. Anyway, seeing Linda is worth the risk."

The remainder of the evening contained a rolling discussion about the air defense system and their experience within it. Several pilots joined in at various times. Most of the pilots simply accepted what they heard over the radio and tended to think of themselves in isolation, or in the context of their section or the squadron. They did not see themselves as part of a larger network. Jonathan shared Brian's curiosity fractionally, but said he was more interested in what the learning process brought with it. Although Brian did not quite see his place in the larger structure, his appreciation for his task and the squadron's mission grew with each lesson, answer, discussion and experience.

Monday, 26.February.1940
No.14 Beauchamp Place
Chelsea, London, England

THE falling snow and half foot accumulation made the city scene seem more like Christmas in Wichita than any time since his arrival in Great Britain. For the first time in many months, he missed his parents and the simplicity of his school days. The nostalgic thoughts lasted until he saw the shiny brass '14' on the jet black door. As he ascended the five steps to the door, his consciousness jumped to what lay ahead on the other side of the door.

The door opened and Anne leapt into Brian's arms in one continuous motion. She wrapped her arms and legs around him in a most unladylike manner nearly causing him to slip on the snow covered porch. She showered him with kisses making him wonder what might have produced the response. He could neither deny nor hide his physical response to Anne's gesture.

"It is good to know you are glad to see me," she said stopping her kissing long enough to speak. "Dear Lord above, it is good to see you. I've been so worried about you."

"Why? Nothing's happened to me."

"The press reports have talked about increasing German activity in the North. Everyone seems to suspect the Germans are going to invade near you. I have just been so worried. I certainly do not want anything to happen to you, after all you are a fine specimen. It would be such a waste . . . ."

"I missed you, too."

"Yeah, I can tell," she giggled mischievously rubbing her still suspended pelvis against him.

"Are we going to do it right here?"

"Now, there's an idea. What do you think?"

"Anne," he protested.

"Oh all right," she responded as she detached herself from him. "Come on in. It is cold out there, and you would probably lose it anyway."

They laughed as she closed the door behind them. He took off his cap, scarf and topcoat. Wrapping an arm around each other's waist, they walked into the living room. Linda Mason stood as they entered.

"I was so wrapped up in my excitement, I forgot to ask you, where is Jonathan?" asked Anne.

"My apologies, Linda. Jonathan's pass got revoked as we were about to leave this morning. The reserve pilot slotted to take his place got sick. The skipper had to hold him back to stand alert. I'm sorry."

"I am sorry as well," said Linda. "I was truly looking forward to see him again."

"Oh Linda," consoled Anne.

"Right, then. Three's a crowd. I should be going."

"Nonsense. Brian, tell her to stay. We can have a private party."

"Anne," protested Brian again immediately thinking back to the night right after the war started when Anne convinced Virginia North to join them. While Brian still remembered that evening with great pleasure, he felt as embarrassed for Linda as he did for himself.

"Relax, Brian. I am not suggesting we all get naked and bonk."

"Anne!"

"It is OK, Brian. It is not such a bad idea, I should think."

"Wait! I can't. I won't. Jonathan is my best friend. I fly with him. I depend on him as he depends on me. I can't do that to him."

"You did it to Jeremy."

"That was different." Brian remembered the feelings he had when he found out Anne Booth was a professional woman and her services had been paid for by Flight Lieutenant Lord Jeremy Morrison, Esq. He rationalized Jeremy's drunken state of unconsciousness, and Virginia was a professional woman like Anne. What would it hurt, he had told himself at the time. This was different although he now wondered whether Linda was a professional as well. Listening to his friend without asking the question outright, Brian knew Jonathan did not think so.

Linda Mason did not wait. The situation was plainly awkward, and she knew only she could relieve the pressure. She stood up. "I'm going to go. Brian, please tell Jonathan, I was sorry he could not make the journey. I look forward to the next time we meet." Brian and Anne walked her to the door. Brian helped her on with her topcoat. She kissed Anne and then Brian on the cheek, and left.

Anne gave no indication that the exchange phased her in the slightest. As they entered the living room, she jumped on him again. "I want you," she whispered.

"Can we talk first?"

"I need you inside me . . . first," she said repeating her gyrations on the porch.

Brian felt the searing heat. "Anne," he protested feebly.

"Please, Brian. Please make me happy." She continued her seductive movement until she was satisfied with his response. "That's my stag," she said proudly as she let go of him, extricated the part of his anatomy she was most interested in and pulled up her skirt revealing her lack of underwear. The sight was more than he could bear.

Brian lowered her to the floor. Without removing a stitch of clothing, Brian gave her what she sought. It was a powerful encounter as many of their passionate moments were. They were both sweating as she approached and he reached the pinnacle. Brian recognized his failure to help his lover over the top. As he caught his breath, he moved slowly to attend to Anne's pleasure.

Anne wrapped her arms around his chest lifting her face to him. "Not now, my sweet," she whispered with her soft, warm breath caressing his face. "I got what I wanted. You can take care of me later. I just want to hold you inside." He pulled his knees up and lifted her off the floor until she sat on him. They embraced. The kisses were long, deep and passionate. As the heat died down, she pulled back just enough to look into his eyes. "What did you want to talk about?"

The subject seemed so distant to him at that moment. He struggled to remember not what he wanted to say, but how to say it. Their current entanglement did not lend itself to other than intimate words between lovers.

Anne must have sensed the difficulty within Brian, but she was not ready to separate. "Come now, Brian. Don't be bashful on me again. It is just the two of us joined at the hip. You know perfectly well there is nothing we can't talk about, so just blurt it out. Get on with it."

"Well . . . I suppose . . . it's just that . . ."

"Brian, please," she said showing him her caring and pleading sapphire blue eyes.

"I've been with another woman."

Anne smiled knowingly and stirred on him as if she was trying to satisfy an intimate itch. "You say it like we are married." She contracted key muscles and moved on him more purposefully. "With whom?"

"Jonathan's sister."

"I hope she is over eighteen." She continued her activity with progress.

"Of course, she is. She's a student at Oxford, medical student as I recall."

"Then, she must have a keen appreciation for anatomy."

"Anne, you are incorrigible."

"Yes, I am and you love it. So, is that all you wanted to tell me?"

"No." Brian hesitated to consider his words as he became aware of his response to Anne's efforts.

"Well?"

"I've learned so much from you. I continue to learn more every time we are together. I do love you, Anne, and I don't mean just physically," he said lightly, anticipating her likely quip. "Rosemary, Jonathan's sister, does not know as much as you do, but she definitely enjoys sex. I tried to resist her flirtations because of my relationship with you, but in the end, I gave into her."

"Good boy. You did the right thing."

"I'm not so sure, but it happened nonetheless."

"And, it's going to happen again real soon, if you don't hurry," she giggled as he returned to full excitement.

"Anyway, to cut to the chase . . . why are British women so much more sexual than American women?"

"Isn't that an interesting question?" Anne thought as she played. "As I recall, before you arrived in England, you had only known one woman or rather girl. It could be, you just got lucky here."

Their movement became more purposeful. "Maybe."

"Maybe it's the long nights of winter. We need something to do, you know. Actually, I think if there is a difference it must be we are simply brought up to recognize sex as a normal human endeavor like eating, drinking or breathing. I can only speak for myself here, but I just like the feel of sex. I have ever since I grew breasts and pubic hair. It makes me feel good and I like that."

"It makes me feel good too, but I just can't imagine my parents doing it, I suppose."

"I know my parents do. My mother and I still talk about it although she has never been comfortable with the subject."

"That's amazing."

"No, my sweet. You are the one who is amazing. I truly suspect you are every woman's dream, and I want to enjoy it again now." Anne began to undress Brian. He took the hint and assisted her out of her clothing. When they were both completely naked among the strewn clothing in the living room, Anne literally pushed him onto his back.

Readjusting her position over him, she moved her hips with precision in every respect stopping occasionally to grind herself against him. She played him like a unique musical instrument responding to the mood, the signs, the rhythm, the urgency of the event. Her full, pendulous breasts functioned like enormous magnets attracting Brian's caress and fondling appreciation. Anne moved with experience to heighten her own pleasure and stimulation. Brian worked hard to withhold himself until the groans and shudders of her climax. The enjoyment of her pleasure amplified his own pleasure. Brian could not comprehend how the sensations could be any better than that moment.

The two lovers lay together on the living room floor enjoying the wet, smooth, slippery texture of the other's curves. Not until their respiration receded toward normal did words return. "I have always thought of sex as the most definitive celebration of life – a clear, undeniable demonstration of what it means to be alive and human," Anne said like an instructor. She stroked and kissed him. "I don't know any American women well enough to know about their sex lives." She paused. "I'm intrigued by your question." She touched his face. "I suppose what you may have experienced is more frankness, more openness. What is Jonathan's sister's name?"

"Rosemary."

"I would presume from your question, Rosemary is not ashamed of her sexuality, either."

Brian chuckled with the memory of those few nights at Carlingon Castle. "I think you could say that." Brian realized and acknowledged to himself the respect he had for Anne Booth. He loved her in a special, maybe even unique, way and looked to her as a person, a woman, with far more intimate experience than himself. He instinctively knew he had a lot more to learn from this exceptional woman. "What's the secret to good sex?"

This time it was Anne's time to chuckle. "Men and women have probably been asking that question for many millennia." She considered how she might or should answer the innocent question. "I suppose I would say it is, sensitive attention to mutual satisfaction. Like we've talked about several times, and you have become quite good at. Good sex is a two-way road."

"I can handle that," as he slowly kissed a path down her body. Gradually, carefully, softly, Brian leaned toward his task of love.

"Oh, yes," Anne cooed like a content kitten. "You certainly can. You have learned the lessons quite well."

Tuesday, 27.February.1940
Headquarters, Fighter Command
Bentley Priory
Stanmore, Middlesex, England

THE RAF car stopped in front of Anne's house slightly before the appointed hour of eight. A familiar face watched with appreciation the good-bye embrace and kiss at the large black door of No.14 Beauchamp Place.

Sergeant James MacDougall waited discreetly until the couple separated before he jumped out of the staff car to open the door for his passenger. "Good morning, sir."

"Hey, James," Brian said automatically remembering the name belonging to the face he had last seen eight months earlier. "Great to see you again."

"Likewise, sir."

Pilot Officer Brian 'Hunter' Drummond took a last long look at his lover standing seductively camouflaged to the casual observer in her morning coat. They blew kisses to each other as he entered the car. It was not until the car turned onto Brompton Road that Sergeant MacDougall commented on the recent scene.

"She's a lovely lady, sir."

"Yes, indeed, James. Far more than I hope you will ever know."

"Sounds like you are in love, if you don't mind me saying so, sir."

"Yes, I think I am."

The journey out of the city to the old monastery upon a hill overlooking London took an hour. The winding narrow streets lined with small shops seemed endless leaving an impression of the city's enormity. The car was stopped and the occupants questioned by suspicious guards several times from the initial fenced gateway. The mature forest thick with conifer trees and rich vegetation masked the clearing surrounding the headquarters complex of Fighter Command.

The white marble faced building appeared more like a stately, country manor for some nobleman of which he had seen numerous examples since his arrival than a military establishment. It certainly bore no resemblance to the green, drab, plain, clapboard structures he had become accustomed to at the various aerodromes he had been stationed. The tall, prominent tower made the main building look more like a cathedral or at least a country church than the center of the Air Defense of Great Britain. As Brian had learned over the last few months, the hub of the wheel lay inconspicuously before him that probably explained the meticulous attention of the guards at various points up to the entrance. The last set of guards protected the specific entrance to the main building.

"Good morning, sir. Please state your business," the sergeant of the guard requested courteously just as the other guards had done three times in the last mile.

Pilot Officer Brian 'Hunter' Drummond produced the letter from AOC-in-C. "I'm supposed to see Group Captain Spencer."

The senior guard carefully scrutinized every mark on the single sheet of paper. "Yes, sir. If you would be so kind to wait here, we shall call Group Captain Spencer's assistant to escort you to his office."

Brian nodded.

The high, domed, painted ceiling of the ante-room further dispelled the common description of the occupants. If it were not for the sea of RAF uniforms filling every space, there would be no way to determine the purpose of the facility. Military buildings were not supposed to have awe inspiring artwork on the walls and standing upon tables and stands. Brian marveled at the splendor of the building and its contents.

A uniformed WAAF sergeant strode confidently into the ante-room and toward Brian. He assumed correctly the average looking, modest built, strong woman was John Spencer's assistant, his secretary. "You must be Pilot Officer Drummond, sir."

"In the flesh," he responded feeling the residual heat from last night.

"Yes, I am quite sure. If you will follow me, sir." She did not wait for an answer, turned and walked back the way she came. Brian hesitated with some surprise at the abruptness, then quickly caught up to her and matched her pace. Her straight back and unswerving hips belied the woman beneath the uniform. They walked halfway down the hall to the right. She knocked on a large door on the right. "Sir . . . Pilot Officer Drummond to see you, sir."

"Show the gentleman in," Group Captain John Spencer said as Brian entered the room. He completed his egress from behind his large desk. "Brian, my boy, it is truly a joy to see you again." The tone of John's words struck Brian with surprise and astonishment as if the older man was equally surprised he was still alive.

"It is good to see you, sir."

"Are you having fun with the Spit?"

"Beyond my wildest imagination, sir. Thank you so very much for your assistance. I will be forever grateful."

"Nonsense, my boy. You did it yourself. Your skills with an aeroplane gained you the seat."

"You may say that, sir, but I am indebted to you, nonetheless."

John Spencer moved as though he were embarrassed by the recognition. "Just promise me one thing, Brian," he said with instant seriousness, "fly like the eagle Malcolm taught you to be."

"Yes, sir."

John motioned to a chair adjacent to his. They talked about friends, the sights, the country and tried to avoid any reference to the war. They both knew without words the thoughts that occupied their innermost consciousness. Neither one could talk of them. One because he knew all too well and the other because he sensed what lay ahead.

"Before we get down to business, Mary and I would like you to join us for supper."

"I must return to my squadron by morning."

"We understand, but I can suggest the overnight Pullman which would get you to Edinburgh by 06:00, tomorrow morning. That should do, shouldn't it?"

"Yes, sir. Then, I would be honored."

"Brilliant. We shall make quick work of the professional, and then have an early supper so you can return to duty with margin."

"Thank you, sir."

"Right, then, we are agreed. Shall we?" John said motioning toward the door.

John led Brian further down the hall to introduce him to their boss, the highly respected and revered, tall, salt and pepper haired, gentleman – Air Chief Marshal Sir Hugh Dowding, Air Officer Commanding-in-Chief, Fighter Command. The presence of the man exceeded the myth and legend. Brian felt history in much the same way he felt it when he saw the Winston Churchill, last August. This was a man of destiny, humble, self-effacing, confident and yet burdened. With all that was going on around them, Brian staggered against the time Dowding had for him, a lowly fighter pilot, one of his 'chicks.' The term felt good. It felt right.

John took the tall, distinctive fighter pilot past the guards into the underground bunker. Brian watched with intense, focused interest as several officers explained the process of gathering the essential information from the network of RDF stations, the observers, fighter squadrons, Coastal Command, the Navy, the Balloon Corps and any number of other sources. They collated the information. They filtered the myriad of conflicting elements until they had sufficient confidence in a target. The decision was passed into the Operations Room for dispensation, and then the essential information went to the appropriate group headquarters and sector control station. Aircraft, sections, squadrons and groups of squadrons were commanded into the sky to do battle.

Activity this day in the Operations Room provided the last piece to the puzzle. The waves of blue ebbed and flowed around and over the enormous map of Great Britain, the English Channel, Northern France and the Low Countries. Every single person in the room, from the map table, to the controllers, to the observation deck above, moved with precision, efficiency and purpose. Only two actions depicted on the entire board kept attracting his eyes.

One of several wing commanders involved with explaining the procedures of the Filter and Operations Rooms leaned toward Brian. "Did you say, young man, you are with Six Oh Nine Squadron?"

"Yes, sir."

"Appears one of your blokes just tallied a Yunkers." The man pointed to Firth of Forth estuary. "Seems Gerry is still trying to drop the Forth Bridge and came up short, again."

"Do you know who got him?"

"Red Section, according to the information we have. It looks as if they are still at it. Raid of ten bombers, best we can tell."

Brian's thoughts instantly shifted to Jonathan. His best friend may have gotten his first kill. Flight Lieutenant 'Sparky' Morrow, the quiet, strong-willed Scotsman, always pushed his section, Flying Sergeant 'Junior' Carrolton and Pilot Officer 'Harness' Kensington. Brian felt the knot in his stomach, the urge every hunter has with the realization he missed an enormous opportunity. He watched the blocks in his squadron's area of operations with singular attention until all the blocks disappeared.

While he could not claim an intimate knowledge, he knew how the system worked. He understood what the system could tell him and what it could not. The questions and answers created a continuous stream of words. The images in his mind drew into distinct focus. He understood the mechanics of the air defense system of which he was a part. He had what he wanted, what he sought.

—

O<small>N</small> the way to the Spencer residence, Brian felt for the first time ready for the struggle he knew lay just over the horizon. Malcolm had told him about the feeling. Now, he could speak of it. He would someday be able to tell his own children or pupils.

The Spencer's owned an elegant brick and wood Victorian house among others of a similar vintage on a hill not far from the famous, boy's school of Harrow. Sergeant MacDougall departed before they reached the front door. Mary Spencer waited for them in the sitting room beside a small but effective fire. A smile washed across her face upon seeing Brian. He was rewarded with a hug and a kiss, European style.

Conversation filled the spaces among the drinks, a simple, but tasteful three course meal, and after-dinner laughter and remembrance. The pleasantries brought an unusual relaxation for Brian, but the mantel clock above the fireplace told him he needed to move along to the train station.

"I'd better go," Brian said being succinct rather than subtle.

"Nonsense, the night is young," Mary jumped in to say.

"She is quite right, Brian. There is plenty of time. The last train out to Edinburgh from King's Cross that will get you there before first call is about 22:30." He paused to consider a continuation of his thought. "I could call the squadron to clear you for another day."

"I appreciate the offer, sir, but . . . . ."

"But, what?" asked Mary with surprising aggressiveness.

"I am very grateful for your assistance. I learned a great deal today, and I think it will help me to be a better fighter pilot." The pause helped him form the words to not offend his benefactor. "Today, I watched my squadron engage an enemy raid and shoot down at least one aircraft. I am a fighter pilot. It is what I do. It is all I've ever dreamed of doing. My curiosity kept me from doing what I am here to do."

The Spencers looked at the young pilot with measures of amazement, appreciation and surprise. "Brian, you will soon have more work than any of us can deal with. Don't be in such a hurry. Enjoy the respite while you can."

The words triggered several questions which Brian inherently knew he could get no answer. Group Captain Spencer obviously knew more than he could tell him and especially in the presence of Mary. This time, it was Mary and Brian who stared at John. Brian wanted to ask, but could not. "I really should get back."

The telephone rang. John left the room to answer the phone. Mary's eyes changed demonstrably losing the edge they seemed to have since the first time he met her in the car after his arrival. Brian thought he had seen the look before. As a reasonably attractive woman, she possessed a mysterious quality that remained beyond his grasp. No words passed between them although he received the message from her. The thought she might desire him shot him with a flush of possibility followed immediately by revulsion flooding from the backdrop of her husband and her age. Mary did not look much older than him, but she was married to a man Malcolm Bainbridge's age, older than his father. Her eyes changed instantly as footsteps announced John's return.

"I am afraid I must run back to headquarters, Mary. Something's happening. 'Stuffy' wants me back in."

"John," she protested, "we have a guest. For God's sake, can't they leave you alone for just one night?"

"There is a war on, Mary."

"I've seen no evidence."

"Maybe Brian can give you some real evidence. He is at the sharp end of this affair." John Spencer turned to Brian. "Would you mind staying? You can keep Mary company. I don't know how long I'll be. I will square it with your commander." Brian nodded his reluctant agreement. "Thank you. I shall be off."

John kissed Mary on the forehead and left.

The walls began to close on Brian. He sensed it. He wanted to leave, to be with his squadron. He could not afford to miss any more patrols. He was here to fly and nothing else. Brian struggled with his search for words and in the end gave into his benefactor's request although deep inside something told him, it was not the best thing to do.

"There really is a war, Missus Spencer," Brian said as the door closed. "We've shot down German bombers near my aerodrome."

"I don't want to talk about it."

"But . . ."

"No, buts. I can think of many other things I would rather discuss with you. Would you like some more wine?" she asked as she went to the kitchen.

"No, thank you," Brian responded with no effect. His thoughts drifted to his conversation with Anne just the previous night. He hoped and prayed he was exaggerating the expression in Mary Spencer's eyes, the tone of her words and the movement of her body. Instinct told him he was not. The trapped animal feelings of Christmas night at Carlingon Castle returned with a vengeance and amplified several fold. Rosemary Kensington was his best friend's sister, but she was a single, adult woman of comparable age free to make her own choices. Brian had convinced himself Jonathan would understand when he found out . . . if he found out. Mary Spencer, on the other hand, could present Brian with a drastically more complex and dangerous situation.

She was married to his benefactor, the man who helped him on his road to achievement of his dream, and she was older which meant she was probably far more resourceful. She was intelligent, confident, strong-willed and decisive . . . all attributes that spelled trouble in big letters. While she was an attractive, mature woman, he did not feel the sensuality he felt with Anne or Rosemary.

"If past history is any measure, my husband will not return tonight."

The alarm bells rang in every corner of his mind. His heart rate jumped up several notches. He felt a greater pounding in his chest than almost any time he had experienced in the air. Awkward, unsure, gawky sensations eroded Brian's confidence as well as his clarity of thought. He wanted to be wrong. All words escaped him.

"Which means, it is just you and me," she purred as she lowered her head without taking her dark, cavernous eyes off him.

His heart began to overspeed like an engine without a propeller to hold it back. He could not say anything to her for fear his senses were wrong and he might offend her. There was only one choice. Brian rose from his chair. "I really must be going. I'm on duty tomorrow."

"No, Brian. Please do not leave. I need the company. I need your company."

"I can't. I simply can't." Brian's flustered words seemed to incite Mary. She reached for his hand as he stood over her.

"Relax, Brian. I will not bite. I just want to talk, to enjoy your presence."

Her words did not relax him. In fact, the effect was quite the opposite. The intonation and meaning behind the words surged toward confirmation of his fears. Brian backed away from her like he would back away from a hungry wolf poised to leap at him.

"I am terribly sorry, Missus Spencer, but I cannot stay."

"Certainly you can."

"No, I really can't. I'm sorry."

"Brian, wait," she commanded as she slowly rose from the sofa leaning forward to allow Brian an unavoidable view of her cleavage. Oh God, he told himself. A strange fear began to fill the gaps in his resolve. "John will not be happy when he finds out you deserted me." The reference could only have been meant for one purpose . . . present Brian with an obstacle.

"He will have to understand."

"I suspect he won't." Brian retrieved his top coat, scarf and hat. "Allow me to be candid and quite frank. I have been attracted to you since the first night I met you."

"Oh God."

"I am a lonely woman."

"Oh God."

"I am a young woman. I love my husband, but he does not have time to take care of my needs."

"Oh, dear God in heaven. Please, Missus Spencer . . . ," Brian pleaded with her, frozen in place, with only one arm in his topcoat.

"I am the one who should be pleading with you. I am the one who wants you, who needs you, who has everything to risk." Brian opened his mouth, but no words came out. "Just relax and enjoy it, Brian. I promise you, I will be most discreet. John will never know."

Brian bolted through the door onto the slippery steps. His rush to put distance between them nearly cost him broken bones or at least serious contusions. He ran as best he could, finding relatively dry lanes in the roadway where traction was adequate. He ran for several blocks not really knowing where he was going. He thought of Anne, of seeking further counsel with her. His watch told him he had less than an hour to make it to King's Cross for the night Pullman to Edinburgh. There were no taxis in the area. He eventually hitched a ride from a friendly lorry driver on a night delivery. Brian made it to the rail station in time. Fortunately for him, his assigned sleeper compartment was empty. However, he did not sleep at all that night as his mind raced through the events of the evening and the possible implications. He had to clear his mind somehow.

―

Tuesday, 5.March.1940
The Admiralty
Whitehall, London, England

COLONEL Stewart 'C' Menzies waited several seconds after the door closed before speaking. "Winston, I think the time has come."

"What do you have?" asked the 66-year-old veteran, statesman, minister, Member of Parliament and now once again, First Lord of the Admiralty.

"We have a number of ULTRA messages as well as various other sources within the last few days which provide proof positive Hitler's intention to invade Denmark and Norway. Preparations, as best we can tell, are well underway. An OKW Enigma message to all commands provisionally sets the execution date for the Ninth of April."

"And so it shall finally begin. The Phony War shall end in a month."

"It appears to be a certainty."

"Do we know his military and political objectives?"

"The best indications are, his military objective is swift subdual of both countries to establish Nazi dominance of the Eastern fringes of the North Sea and lines of communication with the iron ore mines of Northern Sweden. We have no direct evidence of his political objectives although it would seem the likely possibility is complete intimidation of Europe including the United Kingdom through a bold flanking maneuver."

Churchill contemplated the stated possibility as well as several others. "Do you have anything to support your hypothesis?"

"There are several references to *Unternehmen Seelöwe*, or Operation Sealion. We suspect it is a planning effort for the invasion of Great Britain."

"Do you think Hitler is that foolish?"

"He is a megalomaniac bloated with his triumph over Poland. The timidity of our response to Poland, I believe has strengthened his egocentric view of omnipotence."

"Well, well, my dear Stewart, your views have become quite focused of late."

"It is my opinion, nonetheless."

"It has been my opinion for nearly a decade."

"My apologies, Winston. We have had some long days recently as we struggled to develop the picture. The image is quite clear with respect to Denmark and Norway. It is still conjecture beyond that point."

"What do you propose we do about this?"

"With evidence this strong, we must inform the War Cabinet, and as we discussed several months ago, I believe it is time to engage the President of the United States. The conditions we agreed upon have been met."

Churchill sank visibly into his chair as he had noticed many times before. Menzies allowed the First Lord to exercise his mind through the reality of the situation and a variety of alternate actions. Stewart trusted Winston's judgment regarding the political currents and his vision of the way forward. Churchill reached for the intercom box on his desk. "Margie, please contact Bill Stephenson immediately and ask him to join me regarding a matter of the utmost urgency." He did not wait for a response. Stewart wondered if she had even gotten the message, but Winston rarely missed a beat. "I will need to discuss this issue with the Prime Minister in private before any presentation to the War Cabinet. I still have my concerns about Lord Halifax. You should prepare for your presentation to the Cabinet at this afternoon's meeting. While we wait for Bill, I will dictate my note to the President."

Menzies nodded. Stewart sketched an outline of a presentation to the War Cabinet that protected the sources while conveying the gravity of information they possessed.

The First Lord called in one of his stenographers and promptly dictated the content he wished.

While neither of the men knew how long it took 'Intrepid' to reach the First Lord's office, they both knew the time was short.

With proper salutations dispatched, Churchill jumped right to the point. "We have evidence of Hitler's impending invasion of Denmark and Norway. I shall have a note to Roosevelt, shortly. Stewart, would you fill in the details for Bill."

The two men absorbed and digested the latest data from ULTRA and other sources. "I trust you agree the conditions have been met for including POTUS within the ULTRA circle."

"Yes, I do," answered Stephenson.

The stenographer knocked, entered and handed a single piece of paper to Churchill. She waited for him to read her work. When the First Lord nodded his consent, she departed and closed the door behind her. Winston reviewed his note several more times before rejoining the others.

"Then, if you both would, please read my draft. While the two of you review my note, I'll get an appointment with the prime minister."

The MI6 Director General took the note from his friend and read it before handing it to Stephenson.

MOST SECRET

27

5-III-40

To: POTUS

    I have asked Intrepid to carry this message personally for reasons that shall become most evident. To avoid any potential compromise, he will brief you privately and personally on a project of the utmost importance and therefore secrecy. As we have discussed previously, I must ask you to share this information with no one. Again, my

> reasons should be quite obvious. Henceforth, it is my intention to share the product of ULTRA with you alone outside His Majesty's Government. Access within HMG is closely held and tightly controlled.
>
> It is with remorse I must inform you of clear, unequivocal evidence of Hitler's plans to invade the sovereign countries of Denmark and Norway in about one month. I share in the assessment that his expected action is a prelude to a much larger objective, namely the conquest of all Europe including Great Britain. My advice to the PM & HMG will be to respect your position regarding involvement. My opinion, as you have heard many times, remains unchanged. Hitler's intentions are toward all of Europe. I believe it is only a matter of time before America becomes embroiled in this dreadful affair.
>
> I trust all is well with you and Eleanor. Warmest regards and wishes to you both, and to the country. We shall endeavour to take every step possible to avoid global conflict.
> Signed:
> Naval Person
>
> **MOST SECRET**

Both men nodded to each other, and then to Winston. Bill Stephenson spoke for both of them. "Well done, as usual, Winston."

"Fine, then, Stewart, you have an appointment to brief the War Cabinet at 16:30 today. I'll meet privately with the Prime Minister at 16:00. Bill, I'd like you to depart as soon as possible for Washington. I think we are all agreed, you are to share with President Roosevelt all we know about Enigma and ULTRA. I will have an appointment for you in two days' time. Answer all his questions, but please convey in your own words the importance of this source and the overwhelming need to protect it."

"As you wish," Stephenson solemnly responded.

"Winston, one more thing," said 'C.' "I think we should send a trusted agent back to Sweden to warn them of what lays ahead. We also have agents in Norway and Denmark for that matter, but we owe the Swedes."

Churchill did not hesitate. "As much as our conscience tells us to warn our friends, in situations like this I am afraid it is not good practice. The source can and most probably will yield information which will save far more lives than we might save in an instant of good intentions. You are correct about the Swedes, though. We must discreetly tell them without divulging the source or even implying the possibility of a high level source. Can we do that?"

"Yes, sir."

"You must impress upon our agent the importance of avoiding any remote reference."

"He will understand."

"Lastly, I would suggest that unless we can accomplish everything we need to complete within the next two weeks we should not take the action. If we have critical agents, then they should be withdrawn as surreptitiously as humanly possible also no later than two weeks hence."

"Rightly so, I should think," commented 'C.' "We will have everything done."

―

Saturday, 9.March.1940
Chartwell Manor
Westerham, Kent, England

WINSTON woke from his near daily, afternoon nap and shuffled out of his small, south-side, studio, personal room into the large study. The latest ship production numbers troubled him, especially in the context of the disturbing successes of the German U-boats on Britain's supply lines from all corners of the Empire. He moved to his large standing desk near the only window, to examine several tables of production data. His thoughts ground through the meaning of the ship production data in the context of the worsening Battle of the Atlantic. He grunted his dissatisfaction.

"Excuse me, sir," came the soft female voice to his consciousness.

Miss Jennifer Scores, one of his many research assistants and loyal stenographers, sat at a small desk near the north door. She kept her head down, ostensibly looking at the pile of transcription papers before her.

"My apologies, Miss Scores. I did not notice you all the way over there."

"Was there something you needed, Mister Churchill?"

"No, not particularly. I am simply trying to find a path through this rather abysmal situation," he said, as his momentary distraction dissipated and his attention returned to his desk and papers.

At just that moment, Clementine entered by the east door. "Damn it, Winston."

Winston raised his left hand, not wanting to be distracted again.

"At least put a robe on. I am quite accustomed to your penchant for nakedness, but there is a lady in the room."

Her husband did not move and mumbled some inaudible words. Clementine waited for Jennifer to glance at her. Missus Churchill motioned with her eyes and head for Jennifer to take her work to another room. Miss Scores clutched a stack of papers and left by the north door, closing the door quietly behind her. She chose one of the leather, over-stuffed chairs to wait for her husband's attention. She knew well not to disturb her husband when he was deep in thought, and he certainly had plenty to be concerned about. The war on the ground quieted after the subjugation of Poland, but the war at sea, Winston's domain, remained intense from the beginning. Winston grunted and grumbled a few more times. Clementine knew her husband was in his element, engaged as she had rarely seen him; however, she worried about what the stress of war would have on him. They were not young as they were in 1915, when he last served as First Lord of the Admiralty.

Winston grunted several more time before he said, "Enough," and modestly pounded his solid oak desk. He turned to face his beloved wife, "What is on you mind, my darling Clemmie?"

Clementine did not speak and only pointed to his personal studio apartment. Winston did not hesitate and returned, wearing his favorite large, scarlet & gold, silk robe.

"Better?" he asked.

"Winston, really, I have no idea how many times I have asked you. You are quite comfortable in all your glory, but it is not fair to the staff. Miss Scores is in her late twenties and I am sure she would prefer not to see your naked form."

"Yes, yes," he barked. "This is my study, not a public library."

"Winston!"

"Any of the staff are welcome to find other employment."

"Winston!" she exclaimed again.

"When thoughts and ideas come to me, I must pursue the moment."

Clementine shook her head as she resigned, and Winston sat in the match chair across from her.

"What brings you to my lair?"

"You mentioned on the telephone yesterday that we had important affairs to tend to."

Winston considered his thoughts. "Yes, I am afraid it is time that we close up Chartwell for the duration of the war," he said with a somber, regretful tone.

"Seriously? This is but a small home in the country. There is nothing military within miles."

"Yes, my dear Clemmie, there is." He paused to connect with her eyes. "Me!"

"Oh, Winnie, you cannot be serious."

"I am afraid so, my darling."

"Is the war really going that badly?"

Winston knew exactly what the intelligence information meant. The survival threatening Battle of the Atlantic was not going well and was going to get considerably worse before it improved or Britain was forced to capitulate for its subjugated survival. Denmark and Norway were next on Hitler's consumption list. The evidence mounted daily that the Netherlands, Belgium and France would experience the new German *Blitzkrieg* in the not too distant future. If that happened, there was little doubt that England would face bombardment and the threat of invasion. Fortunately, His Majesty's Government bolstered the British Expeditionary Force in France to prevent such an occurrence. He wanted to protect the manor house from being a target for German bombers, if they ever came.

"Let it suffice to say, we have reason for such drastic action and my attention must be in London at the Admiralty. The Battle of the Atlantic and these damnable U-boats demand all hands on deck."

"What about the staff?"

"Most will have to be furloughed, I'm afraid."

"Oh Winnie."

"I know, but I cannot see an alternative. Anything that gives the Germans an impression of activity or occupancy will only invite their attention, and it is not the kind of attention any of us want."

"How are earth would they know. We live in the country. We live outside a tiny village. Only our neighbors are out here."

Winston kept eye contact with Clementine and remained emotionless and expressionless as he considered what and how much he should say to his wife. "I am sure you recall the British Union of Fascists, our version of those damnable Nawzee black shirts. MI5 has turned up most of them, but probably not all. With MI5's assistance, Special Branch has captured or eliminated dozens of Nawzee spies. The *Luftwaffe* flies very high altitude reconnaissance flights, and we have no way to know what they might see."

"That does not sound good."

"No, it is not, which is why we must close our beloved home for the duration."

Clementine Ogilvy Spencer-Churchill, née Hozier, knew what her husband's words meant. "Can we transfer any of them?"

"Clemmie, I am at the Admiralty, not no.10."

"I understand. I will take care of everything. I will try to maintain contact with all of them."

"Thank you, my dear. I am convinced this is essential."

Wednesday, 9.April.1940
House of Commons
Westminster, London, England

THE Member of Parliament for Epping, First Lord of the Admiralty, walked proudly although with a heavy heart the three-quarter mile distance down Parliament Street from Admiralty House in Whitehall to the Member's Entrance of Westminster Palace. The cool spring day and the damp of a rain just past gave him an excuse to enjoy the fresh air. He also enjoyed the greetings, comments and encouragement of citizens as they passed. Churchill knew he was one of the most recognizable faces in Great Britain. Although the world events left little to rejoice, Winston drew a measure of inner satisfaction that popular opinion had changed substantially in the last year. At least people were listening even if it was years too late. Winston stood the furthest from an 'I told you so' mentality. He chose to only look forward, to find a solution, to find an end to this monstrous situation.

The journey through the corridors of Westminster Palace reflected the interest of the common man. Members from various parties took the opportunity to ask, listen, discuss and argue the impact of the latest Nazi insult to mankind. Churchill endured the endless debate with recognition of its necessity, but his mind and intermittent thought jumped rapidly to the various military actions currently underway. Although he could tell no one, the most satisfying element, if not the only satisfying item, in the current situation was accuracy of ULTRA. While the source could not be called bountiful, nor could it fill in all the blanks, the Enigma device and the ingenuity of the GCCS experts gave them a month's warning. The product of ULTRA had to be protected at nearly all costs. The value exceeded anyone's wildest dreams this morning at 05:00 as the *Wehrmacht* crossed the border into Denmark.

Prime Minister Chamberlain, looking gaunt, fatigued and deteriorated from events around him and the ravages of ill health, walked toward the small group of MPs. He motioned for Winston to extricate himself which he did. "Winston, as we discussed this morning, I'll make my remarks on the situation in Denmark and Norway to open the session. Clement Attlee has graciously waived his time to you. I know where your heart lies and I am immensely grateful for your support. I would simply ask you to keep your remarks succinct, so we can move quickly to the debate, if there is to be one."

"As you wish, Prime Minister," Churchill responded solemnly and with respect. Chamberlain nodded his acknowledgment and departed for the Member's Lounge.

Even though Winston had had serious, occasionally volatile, exchanges with his party's leader, the importance of the position he held vastly outweighed any past or present disagreements. Winston's sense of honor and respect dictated his actions without the slightest resentment or resistance. It was simply as things should be.

They waited for the Commons Chamber to fill. Churchill sat on the front bench of the ruling Conservative Party to the right of the Speaker's chair. As the Speaker called the session to order, the chamber fell silent. The Speaker called upon the Prime Minister to deliver his report to the House. Chamberlain stood before the Prime Minister's podium. He looked as if he might keel over at any moment. He hesitated either to gather his strength or his thoughts since every eye rested upon him, and every mind waited for his words.

"Honorable Gentlemen of the House," he began, "this is yet another sad day in our history and the history of Europe," Chamberlain stopped and rubbed his forehead. "As I know you are aware, the military forces of Nazi Germany, at dawn this morning, invaded the sovereign territory of the Kingdoms of Denmark and Norway. Denmark has been virtually overwhelmed in less than a day with only minimal resistance. The grossly outnumbered armed forces of Norway have undertaken a spirited resistance," the Prime Minister paused to take several deep breaths as if he were trying to gather the last of his remaining strength. "His Majesty's Government has begun the process of redeploying units of the British Expeditionary Force to assist the Kingdom of Norway in her defense." The House remained uncharacteristically silent as Prime Minister Chamberlain continued his report of the events of the day.

There was no pleasure taken by Winston Churchill or the few of his supporters in the House at being precisely correct regarding Hitler and his intentions. Churchill sensed the mood of the Members changing minute by minute as he had since the war began. The tide of British determination to resist the dark forces unleashed upon the Continent waxed toward the inevitability of cataclysmic confrontation. Churchill also sensed the increasing pairs of eyes resting or turning toward him. Inside the venerable gentleman of English politics and scion of one of Great Britain's most notable families, the distinctive scents of destiny wafted past him. His time upon the peak was drawing inexorably nearer.

Chamberlain concluded his grim report amid rumblings of dissatisfaction with his leadership. Everyone knew the discontent was not sufficient to topple the government; however, the zeal of his vociferous supporters was essentially undetectable. Many in the House now recognized Hitler's ambitions would not be sated with his conquest of the diminutive Denmark or the rugged Norway. Many felt the noose tightening, and they did not appreciate the sensation.

After letting the impromptu verbal jabs flow for a minute or so, the Labor leader, Clement Attlee, stood to the opposition podium. "I yield my time to the Honorable Gentleman, the Member from Epping, the First Lord of the Admiralty," he said matter of factly and sat down. The House fell silent instantly in anticipation.

Churchill stood from his position on the Government bench, three places down from the Prime Minister. He clasped his hands behind his back, stood as straight as his rounded frame would allow, lowered his forehead to scan the full House, and then began his speech. "The Prime Minister said it was a sad day and that indeed is true. But, it seems to me there is another note which may be present at this moment. There is a feeling of thankfulness that if these trials were to come upon our island, there is a generation of Britons here now ready to prove that it is not unworthy of those great men, the fathers of our land. There is no question of fighting for Danzig or fighting for Poland. We are fighting to save the world from a pestilence of Nawzee tyranny and in defense of all that is most sacred to man. This is no war for imperial aggrandizement or material gain. It is a war pure in its inherent quality, a war to establish on impregnable rocks the right of the individual. It is a war to establish and revive the stature of man."

The uncharacteristically short speech from the House's most eloquent speaker left the seasoned and demanding audience gasping for more. They all knew Churchill would find the positive in an otherwise dismally dark situation. He did not disappoint them. These hardened men of politics wanted more encouragement. They wanted more justification for looking forward and upward.

The remaining discussion in the House of Commons this particular afternoon and evening centered on the intentions of His Majesty's Government to wage war as best it could without jeopardizing the defenses of France or depleting the defenses of the Home Islands. The continuing efforts toward mobilization of the nation's mature youth as well as the nearly dormant defense industry attracted considerable attention. The interests and concerns of the House of Commons no longer centered upon nor were preoccupied by the fruitless and blind efforts to preserve peace. An unmistakable air of desperation possessed the House. They all knew they now struggled to regain the lost years of appeasement.

About the Author
Cap Parlier

—

Cap and his wife, Jeanne, live on the Great Plains of Kansas, along with two dogs and a cat. Their four children have begun their families. He is a graduate of the U.S. Naval Academy, a retired Marine aviator, Vietnam veteran and experimental test pilot. Following his military service, Cap served in several positions in the aerospace industry. Cap has retired from the corporate world and can now fully indulge his passion for the story. Cap has numerous other projects completed and in the works including screenplays, historical novels and a couple of history books.

—

Interested readers may wish to visit his website at:
http://www.Partlier.com

At this website you will find Cap's essays and other items, or you can subscribe to his weekly Blog: *"Update from the Heartland."*

Cap can be reached at: Cap@SaintGaudensPress.com

Books by Cap Parlier

Anod series
The Phoenix Seduction (1995)
Anod's Seduction (2004) [reprint of The Phoenix Seduction]
Anod's Redemption (2004)

Sacrifice (2000)
The Clarity of Hindsight (2016)

To So Few series
To So Few – In the Beginning (2013)
To So Few – The Prelude (2014)
To So Few – Explosion (2015)
To So Few – The Trial (2016)

and with Kevin E. Ready:
TWA 800 - Accident or Incident? (1998)

Coming soon from Cap Parlier, **To So Few – The Verdict**, the fifth book of the series novel of flight and a warrior's life.

Web address: http://www.SaintGaudensPress.com

Visit Cap Parlier's Web Site at: http://www.Parlier.com

SAINT GAUDENS

Books by Cap Parlier

Anod series
The Phoenix Seduction (1995)
Anod's Seduction (2004) [reprint of The Phoenix Seduction]
Anod's Redemption (2004)

Sacrifice (2000)
The Clarity of Hindsight (2016)

To So Few series
To So Few – In the Beginning (2013)
To So Few – The Prelude (2014)
To So Few – Explosion (2015)
To So Few – The Trial (2016)

and with Kevin E. Ready:
TWA 800 - Accident or Incident? (1998)

Coming soon from Cap Parlier, **To So Few – The Verdict**, the fifth book of the series novel of flight and a warrior's life.

Web address: http://www.SaintGaudensPress.com

Visit Cap Parlier's Web Site at: http://www.Parlier.com

SAINT GAUDENS